Science on the Air

Science on the Air

*Popularizers and Personalities on
Radio and Early Television*

Marcel Chotkowski LaFollette

The University of Chicago Press

CHICAGO AND LONDON

MARCEL CHOTKOWSKI LAFOLLETTE is an independent historian who has taught at the Johns Hopkins University, George Washington University, and the Massachusetts Institute of Technology. She is the author of several books, including *Making Science Our Own: Public Images of Science, 1910–1955*, also published by the University of Chicago Press, and *Reframing Scopes: Journalists, Scientists, and Lost Photographs from the Trial of the Century*.

The University of Chicago Press, Chicago 60637
The University of Chicago Press, Ltd., London
© 2008 by The University of Chicago
All rights reserved. Published 2008
Printed in the United States of America

17 16 15 14 13 12 11 10 09 08   1 2 3 4 5

ISBN-13: 978-0-226-46759-7 (cloth)
ISBN-10: 0-226-46759-7 (cloth)

Library of Congress Cataloging-in-Publication Data

LaFollette, Marcel Chotkowski.
    Science on the air : popularizers and personalities on radio and early television / Marcel Chotkowski LaFollette.
        p.   cm.
    Includes bibliographical references and index.
    ISBN-13: 978-0-226-46759-7 (cloth : alk. paper)
    ISBN-10: 0-226-46759-7 (cloth : alk. paper)
    1. Science news—United States—History.   2. Communication in science—United States—History.   3. Radio—United States—History.   4. Television—United States—History.   I. Title.
    Q225.L343   2008
    509.73′0904—dc22
                                                                2007049905

In memory of my grandfather,

CARL ALBERT BISCHOF,

who taught me to read and to listen

Contents

Figures

PROLOGUE

———— ✳ ————

COME join "expeditions to the frontiers of research!" Listen as "eminent men of science tell of their own achievements!"

Watson Davis was not beckoning passersby to a Chautauqua tent but welcoming listeners to a wireless lecture hall, a radio show called *Adventures in Science*, of which he was the host. For a brief while in the 1920s and 1930s, thanks to the magic of radio, Americans could regularly tune in to science. Perched in urban parlors or farmhouse kitchens, ordinary people all over the country could hear some of the world's most illustrious scientific authorities expound on astronomy or atomic physics, describe the biology of antelopes or the habitats of ants, explain the chemistry of nylon or the composition of cosmic rays.

During the twentieth century, scientists claimed intellectual prominence in American culture, and their work came to hold unprecedented significance for the survival of life on earth. Biologists, chemists, and physicists probed the secrets of cells, molecules, and atoms; archeologists, anthropologists, and psychologists dissected the remnants of past societies and the implications of the human condition. The technological fruits of scientific understanding—automobiles, motion pictures, even radio itself—became essential and commonplace. Yet at the same time, the way in which most people learned about science outside schoolrooms and textbooks—that is, through the mass communications media—became increasingly shaped by entertainment values. In the competitive world of commercial broadcasting, science was dramatized, personalized, and eventually marginalized. Despite the efforts of dedicated science popularizers who sought to use radio (and later television) to convey the joys of scientific endeavor and the insights of the latest research, serious science faded from the airwaves.

Science on the Air chronicles the efforts of those popularizers, from 1923 until the mid-1950s, as they negotiated for time on the air. Broadcasting, they soon discovered, was not like print, where publication outlets were seemingly unlimited. Newspaper and magazine articles and popular books about science could be written and disseminated with relative ease, and no single group of gatekeepers could stymie all print popularization. With radio, early promises of boundless public access evaporated in the face of commercial pressures. The exponential expansion of radio's audience in the 1920s, from thousands to millions, and broadcasters' recognition that each listener represented a potential consumer, quickly turned radio into a business. The microphone afforded access to a valuable market, and broadcasters controlled access to the microphone. Programs became the glue that bound together the advertisements. Moreover, the medium's very nature—and, in the United States, compliant government regulatory policy—restricted who could broadcast, a fact that encouraged competitive struggles for a limited supply of airtime.

In 1923, when the Smithsonian Institution initiated a series of "Radio Talks" on science, its prime-time presentations had an engaging amateurishness. The programs spoke to diverse, inclusive audiences, a circumstance that offered extraordinary opportunities for civic education. Access did not depend on donning the latest fashions or being of the "right" gender, race, or social class. You could listen in your overalls or while cooking dinner, be eight or eighty, live at the edge of a forest or on the fourth floor of a Chicago walk-up; you could listen to a neighbor's set or perhaps build your own; you did not even have to know how to read. Radio's science in 1923 may often have been presented by elite amateurs but its impact was inadvertently egalitarian.

The first broadcast popularizers had somewhat naive assumptions about their listeners' interests, but their enthusiasm for science could be infectious. Entrepreneurs like biologist Austin H. Clark, nature writer Thornton W. Burgess, astronomer Harlow Shapley, and professors on university campuses across the United States insisted that guest speakers adopt a formal tone. Lecture topics were selected for educational value, guests were chosen for expertise rather than "radio personality," and Americans tuned in.

By the late 1920s, radio productions had become institutionalized. From time to time, over the next decade, scientific organizations would become involved in the medium, but they never committed the money necessary to create high-quality programs that could compete with comedians, dance bands, or detective dramas. Only one venture, the nonprofit science news organization called Science Service, maintained a steady commitment to radio over the next few decades, cooperating with commercial broadcasters in developing science programs.

By the mid-1930s the networks were increasingly pressuring popularizers to emphasize scientific personalities over facts and excitement over education. By then, Science Service journalist Watson Davis had been introducing or interviewing scientists on the air for over a decade. He had become quite adroit at drawing his listeners' attention away from daily routines or heat waves or the latest news from Europe. The world stood on the brink of war, but *Adventures in Science* consistently offered reassuring snippets of expert knowledge amid promises of science-fueled economic progress.

The program for the night of July 17, 1939, included a typical array of vignettes, each topic chosen to entertain as well as inform. The "Red Planet" would make a close approach to earth the following week. Should the audience "expect a visitation of Men from Mars"? the announcer asked. Davis quickly replied no. His CBS colleague Orson Welles had caused enough panic the previous Halloween Eve with a vivid adaptation of *The War of the Worlds*. Nevertheless, Davis, like countless other popularizers, could not resist adding a few suggestive references to the planet's "canals" before returning to life on earth. As Davis interviewed two government wildlife biologists, the audience was encouraged to dream about Yellowstone's vast wilderness, Alaskan brown bears in Glacier Bay, and the rare nene bird in Hawaii.

The best shows added just enough excitement to spark the imagination, yet not enough to shatter science's dignity. While Davis's scripted interviews mingled scientific fact with occasional touches of drama, other series in the late 1930s, like the Smithsonian Institution's *The World Is Yours* and DuPont's *Cavalcade of America*, were merrily fictionalizing science. Without at least a nod to entertainment, programs could not remain on the air. Soon, though, a nod was not enough, and the decorous voices for science grew faint.

Radio would seem to have been the ideal medium for describing the latest scientific work, an ideal way for farmers in Nebraska or teachers in New Mexico to hear about astronomers, physicists, and biologists. In radio's heyday, however, science programs formed only a minuscule portion of the overall schedule, compared to the volume of soap operas, comedy and adventure shows, sports, and music broadcast on national networks. Why was this the case? At the time, every research field shimmered with intellectual energy. Popular magazines and newspapers in the 1930s featured ever more science content. In 1945 the atomic bomb gave science an unparalleled immediacy for world survival. By the 1950s government funding of research had zoomed into multiple millions in every developed country and every currency, and scientists hovered next to presidents and prime ministers. During a brief period after World War II, radio interview shows and documentaries had played an important role in facilitating public debate on atomic energy, but at no time were the airwaves brimming with science.

The explanation involves the attitudes of the sources (scientists) toward popularization and of broadcasters toward science. Although a few successful, accomplished scientists did become involved in radio (and eventually in early television) for reasons linked to ambition, money, and idealism, the scientific community as a whole remained uneasy and suspicious of the mass media. Moreover, neither the most famous and articulate Nobel laureates nor the most prestigious associations could gain access to the airwaves without the consent and cooperation of individual stations or national networks. And those businesses had become convinced that science would not sell.

This history of science broadcasting progresses through the experimentation of the 1920s, the social concerns of the Great Depression, the worsening European situation, and global conflict and into the postwar debate over government censorship of science and the opening decade of television. Understanding broadcasting's history is, to paraphrase Marshall McLuhan, essential to understanding what happened to science's public messages across that time. The audiences were large and varied; the content's potential impact on them, substantial. As such, the story of broadcast popularization provides a cautionary tale for future opportunities to popularize science through whatever may be the next popular communications medium. If we value a scientifically literate society, if we want people to reject tomfoolery disguised as science, to ask reasonable questions about research parameters, and to understand why knowing more about science will help them make wiser consumer, medical, economic, and political decisions, then we must consider how they will learn about science once they leave formal education behind. How will information about science reach general audiences in the future? Who will take responsibility for delivering science to the public? What constraints might new media formats and their owners or managers impose—and with what consequences for the message?

Settle down, then, to read about energetic, creative people who tackled a worthy goal—to explain to millions of people about cell biology, photosynthesis, and treatments for pellagra, about gas turbines and dwarf stars, about how hot weather affects the human body and why ducks fly south in the fall. Through their correspondence, notes, and scripts, this book documents the challenges those enterprising popularizers faced, celebrates their successes, analyzes their failures, and offers potent lessons for civic education about science today.

CHAPTER ONE

———— ✳ ————

Tuxedos and Microphones

This is merely a matter of salesmanship. The station acquires a higher appreciation of science if science appears in tuxedos than if it appears in soft shirts and patched clothes. . . . Filtering down to a part of our audience, it has an elevating influence on them. AUSTIN HOBART CLARK, 1925[1]

ON October 3, 1923, the head of the Smithsonian Institution had just returned from fieldwork in the Canadian Rockies when his clerk took a telephone message from the manager of a new Washington, D.C., radio station. "Radio Corporation of America desires the Secretary to speak for broadcasting some time between the 20th and 30th of this month, from 8 to 10 o'clock in the evening on any subject," the clerk's note read.[2] Had Charles D. Walcott accepted the invitation, his appearance would probably have been a one-time event. The geoscientist was neither a dynamic nor a comfortable public speaker. After many months away, he had work to do in the "Castle," the redbrick Victorian-era headquarters for the museum complex. It was easy to bump this invitation down the line.

"I do not feel that I could undertake this personally, as my voice is not adapted to it," Secretary Walcott wrote to Austin Hobart Clark, a middle-aged curator in the Smithsonian's U.S. National Museum, "but it occurred to me that perhaps you might be willing to give a short talk and to make the necessary arrangements with the Company as to time, place, etc."[3] Clark was *very* willing. Here, finally, might be an opportunity to make his mark.

If the request had been handed off to some other Smithsonian scientist, the result might have been less eventful. Clark, an expert on marine invertebrates like starfish and sea urchins, knew little about radio broadcasting, but he was smart, ambitious, and exuberantly confident in his intellectual agility. He also had an authoritative and patrician New England accent—an authentic version of radio's stereotypical voice of the 1920s.[4] After his first successful appearance, Clark began, at the encouragement and invitation of the station, to arrange more "Smithsonian Radio Talks," inviting speakers from within his institution's various bureaus.

Those first programs now seem stark and unsophisticated. Succinct fifteen-minute talks delivered by ichthyologists, astronomers, and entomologists. No jokes, no banter, no clever sound effects. Clark's broadcasts found an audience, however, and they represent one of the first sustained efforts in the United States to use radio to reach the public with science, the first wave in steadily expanding popularization efforts. The circumstances surrounding these pioneering broadcasts illuminate an important moment in the history of science and of American life. From the 1920s through the 1940s, the scientific community became more concerned about scripting its public image for the sake of increased funding and political support, and therefore became more engaged in popularization. The extent to which prestigious researchers in the United States cooperated in and supported ventures involving radio—a medium they regarded as intrinsically sensationalistic—demonstrated begrudging acceptance of the need for popularization.

This preoccupation with image coincided with increasing audience interest in science. After World War I, provocative changes were occurring in "the patterns and channels and beaten ruts" of conventional thought.[5] As each scientific insight or discovery reshaped the intellectual landscape, it also wiggled the lens of public understanding. From theories of relativity to theories about human psychology, from animal physiology to plant pathology to Tutankhamen's tomb, excitement surrounded popular discussion of science. Ordinary Americans were eager to learn more about scientists' accomplishments, about exotic aspects of the natural world revealed during expeditions, and about the science underpinning consumer technologies like radio. Topics "which in previous generations would have got into print only in the form of treatises that would have reposed on the shelves," historian Mark Sullivan observed about his own times, "were now among the books most in demand at public libraries."[6] Newspapers and popular magazines like *Collier's Weekly* and the *Saturday Evening Post* played critical roles in satisfying the public's desire for quick and comprehensible explanations. In 1923 radio offered another tantalizing outlet for communicating about science.

Using radio to reach that potential audience proved to be as challenging as it was tempting. Within two years, Clark was declaring:

> One never knows, of course, how long this will last. But all of those who have talked over the radio . . . have acquired a wholly new concept of popular science, and the training they have received will not soon be forgotten.[7]

What was that "new concept"? What did scientists like Clark believe they were accomplishing when they cooperated with commercial broadcasters? And did they succeed? As communications analyst Anthony Smith emphasizes, the "essence" of broadcasting is that editorial power over content is "fused with the ownership of the means of disseminating the product."[8] This innate condition placed scientists—accustomed to controlling their own communications—at a disadvantage, especially when power over all aspects of broadcasting, from scheduling to scripts to access, became consolidated in the United States within a few corporations that owned the national networks. The dream of reaching enormous audiences was alluring. For the scientists and their organizations, the question eventually became whether they were willing to pay the price.

BIG DREAMS

In the fall of 1923, radio dangled in the public imagination like a shiny new toy. Wireless transmission of sound to mass, undifferentiated audiences represented an astonishing prospect, one that frequently prompted predictions of national unity, universal education, and international cooperation. As one commentator observed, radio's success stimulated the dream that society "shall wake up one bright morning with an international consciousness . . . and the dawn of mutual understanding and world peace will have come."[9] To the individual listener, intent on an evening's entertainment, radio simply seemed magical—you tuned in a receiving set, which you might even have assembled yourself, and soon a disembodied voice or musical performance filled the living room.

And ever more living rooms contained sets. Diffusion of radio technology had progressed steadily since the war ended. Station KDKA in Pittsburgh had broadcast the first scheduled programming in 1920; by January 1, 1922, thirty broadcasting stations were licensed to operate in the United States; fourteen months later, there were 556.[10] "The rapidity with which the radio craze swept the country between 1920 and 1924," historian Susan J. Douglas explains, "prompted analogies to tidal waves and highly contagious fevers."[11]

In 1922 science journalist Watson Davis compared "the story of radio" to "a rush to a new gold field":

> Manufacturers of apparatus saw the possibilities of selling outfits if they broad-casted music and entertainment. . . . Radio became an industry. It ceased to be only a plaything for scientifically inclined boys. It came out of the laboratory into the world.[12]

That year, U.S. sales of sets and equipment totaled $60 million, and the net profits of just one manufacturer, RCA, exceeded $3 million, up from $400,000 in the previous year.[13] The price of sets had begun to come within the reach of more consumers, and the equipment was simpler to construct. All over the country, Americans were huddling around newly built sets or seeking better reception in the attic:

> The little room upstairs, once sacred to the storing of books, linen and ancient relics, is coming into its own as the most popular room in many households. Cobwebs have been brushed from the old attic and family clans are gathering there to listen in on the radio voices of the air.[14]

Radio fever was sweeping the East Coast. According to one government estimate, there were between five thousand and eight thousand radio sets, and somewhere between sixty-five thousand and seventy-five thousand regular listeners, in the District of Columbia alone in 1923.[15] That October, as Austin Clark was arranging his first programs, station WEAF in New York City canvassed its listeners, asking them to return postcards with information about their occupations, the other consumer technologies they owned, and the leisure activities they enjoyed.[16] Over twenty-six hundred New Yorkers responded. They were listening in via 409 sets, although only 250 of the households had telephones and 332 had phonographs. The majority of these WEAF listeners were professionals (for instance, accountants and physicians) or worked in some technical or engineering trade (such as telephone installers and telegraph operators), yet there were also many post office clerks, firemen, and actors. WEAF's handwritten tabulations provide a snapshot of an emergent "radio democracy," of the users of an egalitarian communications medium that was becoming available to all. As the price of sets dropped and rural sales increased, the potential audience for science expanded beyond literate, educated, middle-class urban residents. You no longer had to be well dressed or of a welcomed class, race, or gender to gain entrance to this electronic lecture hall. You only needed to listen in.

In 1923 "listening in" did require patience. There was only a minimum schedule and much static, even in major metropolitan areas. News and music programs were available regularly to New York City residents, for example, but true national broadcasting (whereby millions of Americans might listen to the same speaker or comedian at the same time) had not yet been realized.[17] Most stations were on the air only a few hours a day. As Susan Douglas explains, the receiving devices were complicated, with multiple dials and adjustments. One "didn't just walk into a shop . . . buy a radio, bring it home, plug it in, and hear orchestral music"; users had to learn "how to listen."[18] Even though it was possible to hear broadcasts from several hundred miles away, early radio involved more promise than reliable delivery.

The skillful radio owner could, however, sample from a wide array of content. The economic model that exists today—radio as an advertiser-supported medium that relies on sophisticated entertainment to attract listeners to the commercials—was still in an embryonic state. Many stations were owned by organizations like churches and universities, and few accepted any form of commercial advertisement. Talent was drawn from within the community. On a Saturday night in December 1923, Washingtonians might listen to local coloratura soprano Elfrida de Roda and violinist Karla Kleibe, and then roll up the rug and dance as the Meyer Davis Band played live from Cafe Le Paradis; or, with luck, they might tune in distant stations to hear sisters Mary and Jane King sing popular songs "in harmony" or Dr. David Wilbur Horn, professor of physics and physical chemistry at the Philadelphia College of Pharmacy and Science, lecture about "Atmospheric Comfort."

To Clark's first guest speakers, broadcasting's vaudevillian context might have seemed foreign, but they would have been quite familiar with the technological process of wireless transmission of sound. Even if they were not physicists or engineers, they would have been reading about radio technology for many years in journals like *Science* or *Scientific Monthly*, and some might have attended demonstrations at conferences and in colleagues' laboratories. This technological comfort level may have encouraged some scientists to perceive radio as merely an extension of the lecture hall, for many of them did not yet own sets themselves or else had only small crystal sets. California Institute of Technology electrical engineering professor R. W. Sorenson confessed in 1922 that, after fifteen years of studying the technology, he had only just become "a radio convert," with a receiving set at home and a regular listening schedule.[19] Secretary Walcott and his wife did not purchase their first radio until a year after the Smithsonian series began and Walcott had spoken on the air.[20] Those who did not own sets may not have realized the extent to which musical performances dominated the airwaves, thereby contributing to naive assumptions about how science should be presented and how it would be received.

Local station managers were at first delighted to broadcast talks by scientists from well-known institutions. The scientists filled out the station schedules without embarrassment and without added cost. If speakers proved popular with listeners, then they would be invited to return.

Commentators in the early 1920s hailed radio as humanity's "Next Great Step Forward," declared it ideal for promoting social progress or "education and happiness and democracy," and suggested that it would help to "civilize" America and unify a nation struggling with waves of immigrants.[21] Understanding of science, many also assumed, would surely be part of that civilizing force.

EXTENDING A COMFORTABLE FORMAT

In late 1918, Secretary Walcott had just resumed "the old custom" of sponsoring popular lectures by Smithsonian staff members as "a good means to interest the people in Washington both in the various activities of the Institution and also scientific work in general."[22] Austin Clark's first programs transferred this same sedate approach, and similar topics and speakers, to the broadcasting studio.

To oversee the live presentations, Walcott had created the Committee on Popular Lectures, headed by biology curator Leonhard Stejneger, National Zoological Park superintendent Ned Hollister, and Bureau of American Ethnology chief J. Walter Fewkes. The committee selected speakers, sent notices to newspapers and local high school teachers, and succeeded in attracting large audiences to the (free) illustrated lectures on Saturday afternoons. Assisted by ample lantern slides, L. O. Howard from the U.S. Department of Agriculture had talked about "Harmful and Beneficial Insects," Fewkes had described "The Indians as Stone Masons," and Secretary Walcott had attempted to personalize geology with a talk about "My Work in the Rockies." When George P. Merrill lectured about meteorites ("these interesting wanderers in space"), the Smithsonian publicity release explained that Merrill would "take for his text the passage of a meteor over Washington" a few weeks earlier, thereby attempting to add relevance and immediacy.[23]

These public lecture programs continued through 1920 and were thereafter extended through a modest national dissemination effort. Twelve talks, written by Smithsonian experts "in a style to be instructive and entertaining to a general audience," were mimeographed and distributed to YMCA facilities throughout the country, where they were read aloud by local speakers.[24] These specially prepared talks described such topics as the physics of the sun, the geology of "natural bridges," and "antiquities of the Bible." Clark had written three of the talks—"Strange Facts in Nature," "Flying Animals," and "Interesting Animals and Birds from East Africa"—and his supervisor had rated them "very entertaining and instructive."[25]

The Smithsonian leadership initially assumed that radio would provide the same appropriate and positive publicity as the lecture hall series. "It seems to me that this offers an opportunity to bring the public in closer touch with the Institution and might, therefore, be worth while," Walcott wrote hopefully to Clark.[26] The scientists constructed the radio talks as if they were just another set of lectures, requiring no unique techniques, language, topics, or emphasis.

And yet the radio audience was different. It was, for one thing, invisible to the speaker, its reactions to the performance disassociated in time and space. This faceless crowd, unknowable to the scientists who spoke, was a type of audience that, historian Russell Lynes points out, had been "inconceivable in the nineteenth century."[27] Most public speakers "do not realize how much they rely on their audiences for enthusiasm," chemist D. H. Killeffer wrote in 1924, and they can be disconcerted by the experience of "addressing an unseen audience whose presence does not make itself known."[28] Success in this new medium would require a reconfiguration of the popularization process, and a new type of voice.

A NEW VOICE

When he first stepped in front of a microphone, forty-two-year-old Austin Hobart Clark had been at the Smithsonian's U.S. National Museum since 1909. After many years of feeling stymied by his senior managers, Clark was becoming hopeful of advancement. In 1920 he had been made a full curator in the Division of Echinoderms, in charge of his own collection of marine invertebrates. The following year, he parlayed friendship with Professor Jules Richard of the Musée Oceanographique in Monaco and an acquaintance with Monaco's Prince Albert I into a role as the prince's "aide-de-camp" and informal press representative during a royal visit to Washington.[29] Something of a polymath, fluent in French and German, and described as "a man of sparkling humor and quick wit," Clark was a convivial member of the Cosmos Club, an exclusive private club that anchored Washington's scientific life.[30] Lucille Quarry Mann, a close friend, recalled that Clark was "so well educated, there wasn't a field that he didn't know."

> One time when the Cosmos Club was having financial difficulties—should they raise the dues or what should they do? Dear old Dr. [L. O.] Howard said, "Well, I suggest we sell the *Encyclopedia Britannica* and put up a notice that Austin Clark will be in the library from two to four every Thursday afternoon."[31]

Clark had acquired a reputation as an eccentric but well-connected and prolific researcher. At Harvard, his B.A. work had focused on ornithology. He was also known throughout his life as a formidable lepidopterist, publishing over sixty-five scientific works on butterflies, including several standard compilations, and discovering several species, and he was honored for his contributions to that field. Early in his career, however, he had switched to marine biology, concentrating on the phylum Echinoderm (which includes starfish, sea urchins, and sea cucumbers), its subphylum Crinoidea ("sea lilies"), and the exotic genus *Peripatus*. By 1923 he was on his way to being one of the world's experts on echinoderms.[32] Clark had published sixteen articles in the previous three years; his first major book, *Animal Life of Sea and Land*, was due out in 1924. He eventually published over 630 articles and books, including exhaustive monographs on the crinoids.

Although based within a public museum, Clark had focused his attention on research and curatorial collecting rather than education or exhibitions. He confessed to one colleague that he had had so little opportunity during his life "to appreciate the views of the average man toward science" that he felt out of touch with their preferences and expectations.[33] It is characteristic of Clark that this circumstance did not inhibit his entry into radio.

Patrician in outlook but never as wealthy as he might have liked, Clark was proud of his family's extensive New England roots ("the Clarks, Austins, and Hobarts being three of our oldest families in New England, and with numberless intermarriages," he once boasted to an English lord[34]). He had benefited from a privileged upbringing in Newton, Massachusetts. His father, Theodore Minot Clark, a well-known architect who taught at MIT, edited one of the premier architectural publications. After graduating from Harvard College, the young Austin decided not to continue with graduate study and worked instead on several scientific expeditions, including, as a newlywed, an eight-month journey to Japan and other sites in the Pacific aboard the U.S. Bureau of Fisheries steamer *Albatross*.[35]

FIGURE 1. Austin Hobart Clark, biologist and curator at the Smithsonian Institution's U.S. National Museum. Clark arranged one of the first regular series of science programs on American radio. Courtesy of Smithsonian Institution Archives.

As a scientist Clark worked assiduously to document everything that was known about some of the most exotic of marine creatures, and he poured almost equal effort into every avocational passion. Clark recorded his views about science, politics, and radio in voluminous correspondence, in which he rarely refrained from offering strong opinions, once observing with astute self-awareness that "I believe that Nature intended me for an atom, because I enjoy reacting."[36]

When Clark first responded to Walcott's request to contact the radio station, he still believed—along with many scientists of his day—that science could best gain publicity through printed media. After giving one radio talk, however, an October 19, 1923, presentation about the Smithsonian Institution itself, Clark realized that broadcasting offered another route. "We have been advised that your talk was received with much interest by your unseen auditors," the station manager reported.[37] Clark quickly brokered a talk by ethnologist J. Walter Fewkes for October 22 and arranged talks by six other local scientists that fall, including F. V. Coville's description of the

National Herbarium and Charles G. Abbot's discussion of the Smithsonian Astrophysical Observatory. These talks elicited many favorable comments from listeners, and the scientists often heard from neighbors and friends who had caught a broadcast. Fewkes forwarded his fan letters to the station manager, saying that "I think it proves without question the desirability of using the wonderful instrument for broadcasting scientific knowledge."[38]

Clark assured his supervisors that the radio talks had been "keeping our name before the public" in positive ways, and praise flowed from the Smithsonian leadership.[39] The institution's official report to its governing board echoed this expectation of benefits from radio broadcasting: "Possibilities in the spread of authentic scientific information through this means are great, since in this way informational talks prepared in an interesting manner go out to an extended audience fully appreciative of what they hear."[40]

Clark had tasted the apple. The following spring, he sought out WRC's station manager and offered to arrange another series.

CLUSTERED CLOSE TO THE MICROPHONE

This time, Clark proposed to attract larger audiences through more "jazzy" topics, suggesting talks titled "The Giants of the Animal World" and "Creatures That Fly and How They Do It." He also offered, in a letter to the station manager, "some items which might be appropriate to the season," for example: "With the coming of spring the average person for a time displays an interest in the returning birds, the reappearing bugs, and other signs of an awakening nature."[41]

In less than two weeks, the second Smithsonian series was under way, scheduled alongside dance bands, fiddlers, religious services, and sporting events. The station chose to lead, on April 9, 1924, with Clark's "Giants" talk (which, the written script shows, promised far more sensationalism than it delivered). The first series had featured Smithsonian staff. Clark now received permission from Secretary Walcott to include speakers from other institutions and government agencies. "It seems to me that informal cooperation of this sort would be of benefit to us in many ways," he wrote.[42] The burden of preparing and presenting talks would thus be distributed, he emphasized, rather than falling on his museum colleagues alone. But he also knew that the project could extend his own political network in Washington.

Science was flourishing then in the capital region. Clark took full advantage of the proximity of private research entities like the Carnegie Institution of Washington and the wealth of experts in federal agencies like the Bureau of Fisheries and the Geological Survey, the entomologists and botanists in the Department of Agriculture, and the astronomers at the Naval Observatory. During the second season (which ran through June 18, 1924), scientists from throughout the Washington area came to describe their own work or expound on topics intended to appeal to general audiences.[43]

Most programs projected a dull dignity, but occasionally Clark promised more excitement, as in "Atmospheric Electricity" or a talk on flying animals. He invited his colleague Charles W. Gilmore to discuss his work on dinosaurs in "Animal Terrors of Past Ages" and, anticipating today's public fascination with charismatic megafauna, asked the chief of the Bureau of Biological Survey to describe "The *Large* Game Animals of North America." That spring, Clark also arranged one of the first talks on American radio by a woman scientist when he invited a visiting Danish invertebrate biologist, Elisabeth Deichmann, to describe "Little Folks in Greenland." In all these programs, Clark faced the perennial challenge to popularizers—how to attract and retain audience interest while preserving science's reputation for reliable knowledge. He thus encouraged some guests to frame their topics dramatically even while arguing to others that scientists must be "conservative" and must "proceed very cautiously" in using radio for popularization, lest public confidence be undermined.[44]

The call to present some unified message about science made sense. Around the country, many other scientific, museum, academic, and government groups were experimenting with broadcasting. By the winter of 1923-1924, Clark had been in touch with scientists at Chicago's Field Museum and the American Museum of Natural History in New York, as well as various universities, offering to share ideas and scripts. He announced that he was developing "a sort of clearinghouse for the scientific programs of these stations"—a boast that reflected his obsession with compiling information about anything and everything.[45] The American Museum of Natural History offered fortnightly radio talks beginning in 1922, sometimes broadcasting directly from its exhibit halls on such topics as "The Economic Value of Insects," "Hunting Beehives with a Bird," or "Sermons in Rocks," but its series did not focus on science alone and there was no single host. Clark and many others criticized the museum's efforts as "deadly dry."[46] Nevertheless, Clark was politic enough to offer to "cooperate" and exchange scripts with museum official George N. Pindar.

At university-owned radio stations, entire science departments would become engaged in developing programs, as with talks presented by the University of Iowa's Department of Zoology.[47] In 1924, when the University of Pittsburgh established a radio studio on campus (operating as a branch of Westinghouse's commercial station KDKA), academic chemists talked about catalysis, dyes, and explosives; biologists discussed evolution and heredity; physicists spoke about relativity; and zoologists described how microscopes work.[48] During 1923 and 1924 science programs were being planned in Cleveland and Buffalo; nature and conservation series were under way at Pennsylvania State University via WPAB; and Percy Ridsdale's *Nature Magazine* began to arrange talks on WCAP in Washington, D.C. In New York City, a section of the American Chemical Society developed talks that outlined chemists' contributions to advances in dyes, petroleum refining, rubber production, and similar topics.[49]

These regional projects supplemented national informational programming already being arranged by federal agencies. In 1920 William A. Wheeler of the U.S. Department of Agriculture had broadcast the first agricultural market reports; by 1922 thirty-five of the thirty-six stations then licensed in the United States were carrying official agricultural market reports, and two years later, the number had grown to 100.[50] The Department of Agriculture eventually established its own Radio Service to create special programs for farmers and other rural residents, and within a few years had begun producing broadcasts for the general public. In 1921 the University of Wisconsin's station (which had been on the air for four years) began scheduled weather announcements; weather forecasts soon became an expected part of radio programming.[51] That same year, a Denver station carried one of the first public health broadcasts (a program sponsored by the American Society for the Control of Cancer) and the U.S. Public Health Service initiated weekly health talks on a naval radio station in Washington, D.C.[52]

The response to educational and scientific efforts like these was gratifying to the arrangers. Letters praising the Smithsonian broadcasts came from listeners as far away as California, Cuba, and Puerto Rico. Clark was convinced that his audience was quite large. He boasted to one prospective guest that only "one in a thousand of those who 'listened in'" had bothered to write and so their "estimated average audience" was probably "about two million."[53] The feedback from listeners—and the assumption that scientists might be reaching millions—provided a powerful incentive to continue.

POMP AND CIRCUMSTANCES

A key to Clark's success may have been his sensitivity to both the needs of the broadcasters and the nervousness and predilections of his guests. He strived to maintain cordial relations with station personnel: "The use of radio for broadcasting scientific information is quite a new development as yet only in the trial stage and therefore dependent upon the good will of the program managers of the stations."[54] But he worked even harder to put his speakers at ease in the studio.

Formal evening wear (such as dinner jackets) would have been de rigueur at many Washington social events in the 1920s. Clark apparently believed that a radio address constituted just such an occasion and asserted that the speaker who was formally dressed would deliver a more dignified talk at the microphone and would make "a most desirable impression on the station personnel who, like most business people, are accustomed to regard scientific men as learned but more or less soiled."[55] "Here in Washington," he wrote,

> I have insisted from the first that all my speakers appear at the station in tuxedos and in a mental attitude to correspond. This is merely a matter of salesmanship. The station acquires a higher appreciation of science if science appears in tuxedos than if it appears in soft shirts and patched clothes. This is reflected in the advertising of our series sent out by the station, and emphasized by the photographs taken of some of the speakers. Filtering down to a part of our audience, it has an elevating influence on them.[56]

Clark, well known as a congenial host, also invited his speakers to dinner, either at his home at 1818 Wyoming Avenue or the Cosmos Club, and then personally escorted the more anxious ones to the station.

Coddling while prodding seems to have been Clark's predominant technique for handling recalcitrant or uncooperative guests. "I always make my speakers show what they can do beforehand, and if they cannot come up to standard they are put off," he wrote. "I always accompany them to the station prepared to galvanize them with a pin or frighten them with a 'SLOWER' sign. The technique of handling assorted radio speakers is quite an art in itself."[57] His advice to the director of the Boston Society of Natural History (who was preparing his own radio series) was simple and direct:

> Don't anticipate difficulties, forestall them. Always have an alternate in the offing for every talk, and if the manuscripts are not produced in time put the procrastinator off. Fortunately nature has created sufficient docile and amenable individuals in every community so that this can always be arranged. . . . Strong-arm tactics are not often required; but the big stick should always be handy.[58]

For the wiry, small-framed Clark, this was tough talk indeed. He clearly enjoyed describing his series as "emphatically an autocratic proposition":

I run mine like a czar. One must be cold-blooded about it as one slip will ruin a series. All friendship ceases at the microphone; only efficiency counts there.[59]

Such bluster was characteristic of Clark. Whenever he felt comfortable with a correspondent, Clark's rhetoric would soar and swirl and then burst with a splash on the second page. Yet, many of his observations about the peculiar personalities of scientists have a ring of truth. He sometimes described stalking potential speakers much as, with butterfly net balanced, he might have pursued an elusive Papilionidae or Pieridae:

I find that the best way to get talks is to make up my list of subjects and then entangle my victims. Sometimes they object; but they can usually be soothed, or flattered, into compliance. If very belligerent they usually calm down in a month or so, and in another month become curious; then they are easily caught. You can clinch matters with chronic hesitators by writing them a cordial letter of thanks for their kindness in agreeing to give the talk, and setting a date. The number impervious to all forms of diplomacy is negligible.

. . . Always approach them with the manner of a suave stock salesman selling worthless stock; with awe in your manner let them understand you are offering a privilege only granted to a few of the select. Never let anyone get the idea that he is assisting you in agreeing to talk.[60]

His explanation of the final stage is outlined with a naturalist's eye for detail:

It usually takes some months to lead anyone to the microphone. Nearly all refuse at first. In a few weeks they get curious. Later they get so very curious that they are as putty in your hands, and you can do anything with them.[61]

Once they agreed, speakers were asked to "bear in mind that they have a more diversified audience than is ever assembled in a single hall" and told that their talks should be "plain and informational rather than oratorical."[62] Still, most scientists inclined toward "lecturing" via the radio. To address this problem, arrangers like Clark required that each guest provide a written version of the talk in advance, then spent considerable time compressing and editing it into prose deemed more palatable to general audiences.

Timing proved to be an especially rocky process. Scientists were accustomed to expounding at length about their research and to including extensive qualifications and caveats in their explanations. Arrangers had to exercise considerable diplomacy to make a speaker conform to radio's schedule. "It seems to me that you have more manuscript than you need" for a ten-minute talk, one coordinator cautioned a scientist who had submitted twice that much. "One has to speak rather slowly to the radio in order to make it perfectly distinct."[63] Speakers were also reminded not to use "any louder voice than is customary in addressing a small audience," because the "loudness is regulated by the man who has charge of the radiation." And, of course, they should "avoid the rustling of paper" as they read from scripts.

Scheduling proved to be another problem. As early as 1924, science programs were being preempted by political coverage and sports events, a conflict in network priorities that continued for many decades. If the Democratic Convention "should happen to be holding forth" on his appointed day, one scientist was told, then "the time table may be upset, since, of course, a politician may take precedence of any mere scientists."[64] The scientists' own routines also provided interruptions. Researchers like Clark tended to leave town in July and August. No one wanted to remain in steamy Washington, of course, but also, the summer break in college teaching by tradition provided opportunities for scientists to go on digs and expeditions, to collect specimens in wildernesses or at sea, or to visit laboratories or libraries abroad. During the 1920s, Clark frequently vacationed in Massachusetts while continuing research at Harvard's Museum of Comparative Zoology. As a result, the Smithsonian broadcasts were suspended for the summer.

MAKING CHOICES

Success brought choices—and more challenges. The third series of Smithsonian talks ran weekly from October 2, 1924, to May 14, 1925, and some of the programs were "rebroadcast" on New York stations WJY and WJZ. For Clark's series, this meant that the original script was mailed to another station, where it would be read on the air by someone else (a local announcer or, preferably, a scientist). By spring 1925, RCA could do coordinated hookups, essentially broadcasting over phone lines, but this was an expensive procedure for programs of modest popularity.

Clark continued to widen his guest list, inviting scientists from Yale University, the U.S. Weather Bureau, and the Navy Department. In January he helped to arrange the first radio transmissions from a meeting of the American Association for the Advancement of Science (AAAS), while also handling the group's press publicity. And again that season, one speaker was a woman— Smithsonian scientist Doris M. Cochran, who described "Lizards and Their Kin" on April 9. Four talks were broadcast simultaneously from Washington and an RCA station in New York, and WCAP broadcast three others. The topics ranged from "How Trees Grow" to "Is the Universe Finite?" and astronomer Henry Norris Russell's account of "The Evolution of the Stars."

Not every scientist relished the opportunity to stand at the microphone. A few, despite Clark's efforts, refused to appear. Clark attempted, for example, to persuade Scottish physiologist J. J. R. Macleod to speak on the radio while the Nobel recipient was visiting Washington for the AAAS meeting, but Macleod declined to "prepare a talk on the history of the discovery of insulin" with the terse dismissal that he thought that the subject "has already obtained sufficient notoriety."[65] Clark did not, of course, care that the topic had received attention elsewhere. He now deliberately attempted to choose topics and speakers with proven popular appeal, and to give talks more enticing titles. During spring 1925 Smithsonian curator Barton A. Bean described various fishes, Department of Agriculture entomologist S. A. Rohwer told "How Some Wasps Live," and, in what would become a common practice, other scientists targeted such nagging summer realities as "Mosquitoes and Other Blood-Sucking Flies" and "Chiggers, Ticks, and Fleas."

Despite ongoing public debate and despite his own work on the subject, Clark assiduously refused to include one topic that would have attracted an audience—evolution. Scientists, he argued, should "avoid controversial subjects" in their radio talks because that might "discredit" science.[66] To maintain public confidence in science, he believed, talks "must be wholly accurate, dignified, and without reference to controversial points."[67] Clark resisted the "tremendous pressure brought to bear to put on talks about evolution," explaining that

> such talks would quite destroy the confidence in our series. We might just as well, under the guise of anthropology, arrange debates between orthodox catholics and followers of Islam.[68]

By 1925, therefore, Clark had a clear purpose in mind, a clear vision of who he thought was listening, and a list of topics he believed would be appropriate. Popular science must be kept "within safe bounds," he wrote, to accomplish the greater goal of protecting science's reputation.[69] Radio, he believed, offered a "wonderful opportunity" to get "the facts of science to the mass of the people, especially the somewhat numerous class who read only with difficulty and thus never get beyond the newspaper headlines."[70]

This vision of purpose was unarguably elitist, reflective of Clark's time and context. He spoke from within racially segregated Washington, from within an institution and profession unconcerned with enabling equity and smugly confident in its intellectual (and social) superiority.[71] Yet here again Clark's honest insights, and his willingness to commit audacious observations to paper, provide a rare glimpse of who was listening to science on early radio. It is a description not revealed in more formal reports or contemporary discussions. He boasted, for example, to one prospective speaker,

Colonel E. Lester Jones of the U.S. Coast and Geodetic Survey, that the enthusiastic responses to the Smithsonian talks had "come to a surprising
extent from laborers and from the colored population, who express themselves as greatly pleased at being enabled to get authentic information which
they can understand on scientific matters, and also with the opportunity of
hearing the voices of important men with whom personal contact is for them
impossible."[72] He made similar assessments of a racially diverse audience in
other letters, and acknowledged that his listeners included ordinary people
with ordinary reactions related to who and what they were. "The [audi-
ence's] appraisal of the comparative interest of talks is wholly a personal
matter of natural predilection, previous training (if any) and prejudice," he
explained. "The supper the listener has just eaten also has an influence."[73]
Clark had previously dismissed the capacity of most of his fellow citizens to
be interested in science. "The general mass of a population in any country is
intellectually inert," he wrote in 1918; "it learns what it is forced to learn, in
and out of school, and no more."[74] With radio, he was seemingly working
to correct that situation while still retaining his elitism.

Through his association with professional broadcasters, Clark learned
that one way to cultivate a loyal audience was by arranging topics to insure
"uniform standards and adequate variety." He also recognized that radio
speakers could not exploit the conventional tricks of the lecture hall to keep
their audiences interested. One could not smile or gesture or demonstrate
an experiment or make sparks fly. "In an ordinary lecture the audience
is half occupied in listening to the speaker and half occupied in watching
him"; if the subject matter is not "diluted," then the "audience will get
tired," whereas in a radio talk the listeners are wholly focused on what the
speaker has to say.[75] The radio talk thus had to "be a catalogue of interesting
information and well presented from start to finish."[76] Clark attempted,
albeit sometimes clumsily, to frame topics in an interesting manner, and he
strived not to belabor the audience with too much material.[77] A title like
"The Cowbird," he explained, is "too grimly prosaic." Better to name it
"Abandoned Bird Babies" and to draw in listeners with an opening sentence
like "Those unfeeling mothers who leave little babies upon the door steps
of prosperous people's houses have their counterparts among the birds."[78]

"GOOD RESULTS"

In May 1925 Clark received gratifying praise from Secretary Walcott, who
congratulated the curator on "the good results that you are getting in
connection with 'Radio Talks.'"[79] To the Smithsonian leadership, Clark's

"good results" appear to have been primarily a matter of selecting the appropriate people to represent the institution, rather than serving large audiences well. The Smithsonian scientists clung to measures of success that reflected how scientists generally chose to judge themselves, measures that were quite different from those used by the broadcast industry. The differences in those yardsticks, which influenced both groups' assumptions about what, how, and by whom scientific material should be delivered on air, shaped the developing relationships between the scientific community and the broadcasters.

To the scientists, success in broadcasting science was to be judged, first and foremost, by how accurate and authentic the material was, and by the stature of the scientists involved, not by the nature of audience reaction or by the number of people who tuned in (or tuned out). The material, they felt, must whenever possible be delivered on the air *by* scientists—that is, in science's own voices. Those first "radio talks" were therefore idealized as abbreviated versions of public lectures and evaluated internally by similar standards.

The audiences engaged in their own idealizations. These encounters via the airwaves gave science a timeliness and familiarity. As Russel Nye and other cultural historians point out, radio provided a new sense of "simultaneity," closing the "time-gap between event and audience, as if the ideas [discussed] were newly conceived."[80] Radio also broke free of popular science's conventional remote, impersonal tone and offered a charming (if deceptive) intimacy, "a person-to-person quality that gave an illusion of private face-to-face communication."[81] In their living rooms, clustered around the wireless, Americans could imagine a perfect scientist, someone who was personable, intelligent, friendly, accessible. Radio, Nye explains, thus "could improve on actuality itself."

To the embryonic broadcasting industry—which controlled access to the air—scientific content was just one of many options, one of many potential routes to profitability. It mattered little to the WRC station manager whether Clark arranged a talk about stars or starfish, or whether a guest said "5.6" or rounded the number to "6," as long as the audience enjoyed the performance. When Secretary Walcott was first asked to appear, the invitation came not from a specific desire to deliver science to the public but as part of a more general effort to recruit talent from the Washington area. The Smithsonian talks nestled respectably between sermons and musical performances and helped to fill the schedule. Once other stations came on the air and radio learned how to "sell" time, competition intensified. Science had to please sufficiently large numbers of listeners—and potential sponsors—to retain access to the microphone.

Clark and the other early scientist-popularizers were playing with a medium far more powerful than they realized, one with a potential impact on mass audiences that was hard to achieve through print. They thought that because they "understood" the technology of transmitters and receivers—that is, the underlying science—they also understood how *broadcasting* worked. Radio, however, was far more than "tubes and wires in a box." It was rapidly becoming a cultural, social, and political force, with enormous economic repercussions through its advertising and with astonishing power to influence public opinion.

Those first few years of science broadcasting, viewed now from a century consumed with twenty-four-hour information-on-demand, have a sweetness and naiveté not unlike the riverbank in *Wind in the Willows* before Toad acquired his infernal machine. In May 1925 Clark initiated contact with a man who made his living anthropomorphizing animals in the great tradition of Kenneth Grahame. A storyteller and successful children's book author, Thornton W. Burgess had also recognized the power of this communications technology and was already adapting it to his pragmatic purposes as well as his social causes. In Burgess, Clark gained both a new friend and a new perspective on radio's potential for communicating about science.

CHAPTER TWO

———— ✳ ————

The Radio Nature League

The very word "science" is repelling to most people. It smacks too much of deep thought and extreme use of the gray matter.

THORNTON W. BURGESS, 1926[1]

IN one of many letters aimed at publicizing his Smithsonian radio series, Austin H. Clark wrote to Thornton W. Burgess, a successful children's book author. Burgess's animal characters and natural history lessons had been enchanting children and their parents for many years, and unbeknownst to Clark, the writer already had his own show, broadcast throughout Massachusetts. As his audiences swelled, the stations had expanded the program from ten minutes to a half hour and had persuaded Burgess to remain on the air during the summer months.

Subsequent exchanges between the two men, preserved in cordial letters over the next twenty years (for a while, almost daily), provide revealing glimpses of what motivated such entrepreneurs to attempt to adapt radio for science popularization. Their correspondence also sheds light on important differences between types of popularizers. Despite many shared attitudes and assumptions, Clark and Burgess disagreed on the main goal of popularization and, especially, on whose interests it should serve.

Clark's entry into radio was directly related to his professional ambition and dreams of institutional advancement. His attitudes epitomized those of the scientific elite—narrowly focused and autocratic, eschewing controversy, and steadfastly arguing that radio must be reserved for education. The ordinary listener, he believed, could not comprehend the significance of scientific knowledge without the assistance of experts, and so experts must control all aspects of science broadcasting. Although he and the Smithsonian paid lip service to goals of social improvement for the masses, they consistently chose popularization as a means to an end, rather than a worthy goal in and of itself.

Burgess embraced more cosmopolitan, democratic, and pragmatic attitudes toward popularization. He assumed that the public was eager to learn about science and nature and that radio could assist that quest for understanding, and he was not averse to employing entertainment to enliven his presentations. He had considerable respect for scientists but did not regard them as the sole proprietors of scientific knowledge. Although Burgess exploited radio to promote his books and to reach out to faithful readers, he also had a passionate commitment to using broadcasting to encourage environmental conservation, at a time when those concerns were emerging in the public consciousness. In the 1920s, American prosperity and unrestricted land development threatened to overrun native wildflowers and nesting grouses. Burgess recognized that radio could be a useful political tool in advancing conservation messages and he began to use his listeners to effect social change and protect vulnerable wildlife.

Burgess thus provided an alternative model for those who were experimenting with radio. When scientists like Harvard astronomer Harlow Shapley and Clark's friend William Mann (head of the Smithsonian's National Zoological Park) became engaged in broadcasting, they continued to celebrate the importance of scientific expertise, but they also heeded the lesson demonstrated by Burgess's success—that a little entertainment could make natural history education palatable as well as nutritious.

BURGESS

Thornton Waldo Burgess had grown up less than a hundred miles from Austin Clark but in far different circumstances.[2] Like Clark, his Massachusetts roots were deep. He was a direct descendent of Thomas Burgess, one of the first settlers of the Cape Cod town of Sandwich in 1637. Family heritage provided little protection, however, when his father died the same year in which Thornton was born, leaving the widow and infant in economic peril. As a young boy, Burgess worked year-round at odd jobs or on local farms, and that resourcefulness and diligence proved useful later in life. After graduating from high school, he attended business college in Boston for a year and then began full-time employment as an assistant bookkeeper. While Clark was vacationing in Europe prior to entering Harvard College in 1899, Burgess was already working in the publishing business in Springfield, Massachusetts. Five years later, Burgess was associate editor of *Good Housekeeping*, completing his first book, and about to marry. *Old Mother West Wind*, the inaugural volume of what would become an immensely popular nature series, appeared in 1910. From then on, Burgess devoted his career to promoting a science-based understanding of nature, eventually publishing over 170 books and fifteen thousand stories and columns, and creating a menagerie of friendly, anthropomorphized characters like "Jerry Muskrat," "Jimmy Skunk," and "Spotty the Turtle."

Although his books were written for youthful audiences, Burgess devoted considerable attention to influencing the attitudes of his fellow citizens toward their natural surroundings. From the beginning, his radio program sought to reach both children and adults. Boy Scouts might be mobilized to collect tent caterpillar egg cases, but Burgess knew that parents were the ones who could change conservation laws and practices. By 1914 his conservation efforts were so well regarded that the Wildlife Protection Fund recognized him with a medal and a banquet in New York. The "Green Meadow Club" competition, created with his longtime illustrator Harrison Cady, focused on setting aside lands for wildlife and bird sanctuaries. Contributions from members of the Radio Nature League sustained a local bird refuge for almost six years.[3] And, with scientist Alfred O. Gross, Burgess helped to document on film the last surviving heath hens on Martha's Vineyard ("It seems a tragedy of tragedies when a species passes out of existence," Burgess wrote after their final trip).[4] The Burgess conservation legacy continues today through an eighteen-acre estate in western Massachusetts, transferred to the Audubon Society after his death.

In his autobiography, Burgess describes an "abrupt" introduction to a radio career that lasted well over a decade. Sometime in 1924, a representative of a local station invited him to read a children's story on the air. Burgess did not yet own a set himself, had never seen a broadcast microphone, and said he perceived radio more as a "scientific plaything" than a powerful platform from which to communicate to enormous audiences.[5] The public's reaction to that appearance was immediate and positive. Burgess returned to read more stories during the children's hour, and soon was asked to give brief talks about birds and animals following each story. By the end of the year, Burgess had expanded the programs, carried on WBZA-Springfield and WBZ-Boston, to focus on nature and conservation.

In January 1925, with a stroke of inspiration (and a keen understanding of his times and audience), Burgess founded the Radio Nature League, an association requiring no dues and asking only that listeners "assist" in wildlife preservation through actions like feeding birds during the winter.[6] Each act of conservation (reported on the honor system) earned a "star" on a member-ship card. Within three weeks of announcement of the League, Burgess had received over five thousand letters, from individuals and entire families in thirty-nine states, Canada, England, and Bermuda, all pledging (per the League's motto) "to do everything possible to preserve and conserve all desirable American Wild Life, including birds, animals, flowers, trees, and other living things; also the natural beauty spots and scenic wonders of all America." By May over sixteen thousand people had enrolled as mem-bers.[7]

The lists of League members, and the letters they wrote, helped Burgess "form a mental picture" of his constituency. He characterized the show as an "open meeting"—"We, speaker and listeners, were just a big gathering of neighbors with common interests, each one free to contribute his or her obser-vation or comment or to ask for information."[8] The audience was diverse, ranging in age from small children to the elderly ("I early discovered that interest in some form of nature lore is practically universal"), and included "farmers, day laborers, businessmen, doctors, lawyers, clergymen, teach-ers . . . and housewives on distant farms and in crowded city flats."[9] Pub-licity in the *Literary Digest* (June 6, 1925) brought nationwide attention, praising the League as a "highly commendable project" and describing such members as a lighthouse keeper on the Maine coast and "a Canadian trapper 250 miles north of Quebec."

"Gradually I realized that I had in my charge an instrument for educa-tion with undreamed of potentialities," Burgess later wrote in his autobiog-raphy. And yet he knew that good intentions would not be enough. Radio popularization required extra effort to hold the audience's attention: "That the programs should be entertaining was of course a first essential."[10]

Because sponsors were reluctant to advertise on educational programs, Burgess received no remuneration from the stations, a consideration that did not initially inhibit the successful author. During the summer of 1925, the *New York Times* had listed Burgess as among the biggest income tax payers in Massachusetts—his payment in 1924 was $14,492.54. "It is true that I get some advertising" from the series, he admitted to Clark, "but I don't want the advertising and don't need it."[11]

PLATFORMS AND PERSONALITIES

When Clark first wrote to Burgess, he asked the writer to contribute to a series of talks being planned with the National Zoological Park. The Smith-

THE BURGESS RADIO NATURE LEAGUE

A NATION-WIDE ORGANIZATION TO PRESERVE WILD LIFE had its inception in a radio address from Springfield, Massachusetts, by Thornton W. Burgess, whose Bedtime Stories about animals have long been the delight of hosts of children and a goodly company of their elders. Westinghouse News Service *Bulletin* (East Pittsburgh) thus tells of the public response to Mr. Burgess' highly commendable project:

"No other agency has met with such success in arousing public interest in wild life and, in the few months since the organization, more than 10,000 men, women and children, from 34 States in

Courtesy of Station WBZ-Boston

THORNTON W. BURGESS, AUTHOR AND NATURALIST

Broadcasting a Radio Nature League story from Westinghouse Station WBZ. The WBZ Nature League, of which Mr. Burgess is the director, now numbers more than 10,000 members scattered throughout the United States and Canada.

FIGURE 2. *The Literary Digest* publicized the founding of the Burgess Radio Nature League in its June 6, 1925, issue. Popular children's book author Thornton Waldo Burgess created the voluntary membership organization to attract listeners to his new radio series on WBZ-Boston. Reproduction from the collections of U.S. Library of Congress.

sonian scientist added a little sugar to the invitation: "Everyone at all interested in nature knows your work—all my five children are enthusiastic about it—and everyone professionally occupied with natural history knows how accurate all your statements are."[12] Burgess responded enthusiastically by return post, thanking Clark for the "kind tribute to my work" and agreeing to speak sometime in the future. He then explained that "possibly unknown to you, I have been on the air with a nature talk every Wednesday since the middle of November."[13]

The two middle-aged men—Clark was then forty-four and Burgess fifty-one—turned out to be kindred souls, sharing similar political values, a fascination with natural history, and the enthusiasm of adventurers in unexplored territory. Each had a record of career accomplishments, took pride in work well done, and was experimenting with a new medium. Each talked of exchanging ideas and "thrashing these things out" when they met in person someday. Each reveled in the fact that they had "much in common" and had "apparently . . . arrived at quite the same conclusions [about radio] independently."[14] Over the next three weeks, they exchanged at least nine letters. They continued to correspond throughout July and August while Clark vacationed on Boston's North Shore. They met several times that summer, and Clark made an appearance on Burgess's program. The letters preserve their dreams, assumptions, and assessments about radio, as well as their differing perspectives on why radio should be used to serve the public interest.

Those views were shaped significantly by how they had arrived at the microphone. Burgess and Clark were not, strictly speaking, part of radio's emerging audience; they had not become involved because they were building sets in their attics or spending hours tuning in far-flung stations. That June Burgess confided that he had perhaps become "a perfect radio crank, but not from the listening-in end [for] I have nothing but a small crystal set which takes in the local station only."[15] Nor were they radio professionals. They were not among the ranks of skilled engineers and managers who were building station operations, creating clearer sounds and smoother broadcasts, drawing up schedules, and making content choices. They were not among the actors, entertainers, or announcers who would become broadcasting's celebrities. They were not the industry leaders and executives figuring out how to assemble viable business connections and investments to create what would soon become the national networks. They were popularizers of natural history, adapting this new technology to reach a wider audience, enthusiastically engaging in public education. Each had just begun to work out how to use a script and a radio microphone to communicate the same ideas he was already accustomed to conveying through publications and formal lectures.

They exchanged ideas and opinions via those first letters much as they might have done over dinner at the Cosmos Club, exhibiting the naive optimism of amateurs. Both men had recognized radio's potential, and both dreamed that it would be used for positive goals. Burgess predicted that "the time is coming when . . . people are going to have their sets for the sake of the really fine things which they will be able to hear on the air."[16]

"We have in the radio," Clark wrote, "a most useful instrumentality for correcting various misconceptions" about natural history, and that makes it all the more important to "maintain a high standard" in one's broadcasts.[17] "Pitfalls there are," Burgess agreed, and he seized on his acquaintance with Clark to check the statements in his nature stories and to answer listeners' questions by forwarding bugs, cocoons, and other bits of nature they submitted. Clark and other Smithsonian scientists did their best to identify every specimen and answer every question. "I hate to bother you with this sort of thing," Burgess would say in apology as he enclosed yet another mysterious insect or seashell, "but when people are interested enough, and when they have used their eyes sufficiently to note a thing of this sort, I like to hand them out a little information if it is a possible thing."[18]

Each had strong opinions about rival broadcasters. Neither seemed to have admired the American Museum of Natural History programs, finding their lack of an "official voice" and poor delivery to be hindrances to positive impact ("... there is plenty of good material there, but... no one with sufficient ability or backbone to organize it for presentation").[19] Clark in particular could assess another man's broadcast potential with brutal honesty, once declaring that a prominent Harvard scientist's "English is normally rather lurid, and I am not sure that he would not explode if he got 'rattled' before the microphone."[20]

They disagreed on whether science should always be presented impersonally or should include more human elements to keep the audience interested. Clark argued that, as the organizer, he should be perceived as aloof from any desire for fame and celebrity (no matter how much he may have actually craved it). "In running the Smithsonian radio talks," he wrote, "I have preferred always to keep myself in the background and to bring others forward."[21] He also believed that science and its popular presentation could and should be divorced from their human context. Burgess, who had had extensive experience as a public lecturer and popular writer, argued for introducing more personality. "I am for any legitimate method of popularizing scientific matter," he told Clark:

> It has to be done, if we would reach the great mass of American people. The very word "science" is repelling to most people. It smacks too much of deep thought and extreme use of the gray matter. Yet the public is eager for a simple understanding of the various branches of science.[22]

To reach radio audiences successfully, Burgess believed, a speaker must project "on the air, just the same as from the platform," in that "you must put your personality out over the footlights."[23]

Whatever their differences, they shared a newly acquired passion. "You and I have certainly got to get together this summer," Burgess wrote a few weeks into their correspondence. "I believe that between us and the two stations . . . we can accomplish a tremendous thing in stimulating interest in various lines of nature."[24]

SUMMER IN BOSTON

On June 29, 1925, Austin Clark loaded his wife, children, and luggage into his car and drove up to Massachusetts to spend the next two months on the shore. They stayed in Manchester, a picturesque town on the rocky coast a few miles north of Boston, a vacation spot favored by the New England elite. Clark planned to journey into town two days a week to continue research on specimens at Harvard's Museum of Comparative Zoology, and he invited Burgess to drive over from Springfield and spend the night with the Clark family. After that meeting, they arranged for Clark to give a radio talk about butterflies on Burgess's WBZ series and met several more times in Boston before summer's end.

Some of that time they spent plotting to establish a network of radio pop-ularizers and what Clark called a "clearing house" for natural history talks. Burgess also persuaded his stations to agree to a second natural history series, coordinated by several Boston institutions. An August 21 meeting at the museum of the Boston Society of Natural History, planned by Burgess and Clark, brought together director Edward Wigglesworth, Harvard University pro-fessor and herpetologist Thomas Barbour (president of the Boston Society), and physician John G. Phillips (a close friend of Clark's). Also engaged in this effort was Clark's cousin, Edward J. Holmes, director of the Boston Art Museum. "These gentlemen will form an assembling agency and clearing house" for radio talks on natural history, "suitable for the younger genera-tion" but "prepared to interest older persons."[25] A committee of scientists would coordinate talks and select suitable speakers for broadcast over WBZ. The scripts of these talks would be shared with Clark for reading on the Smithsonian program in Washington, with proper credit to the original speaker, and Clark would also share his speakers' talks with the Boston group. This arrangement continued for at least two years, until the Boston Society transferred its talks to WEEI, a station owned by the *Boston Transcript*—a ven-ue that provided more publicity "where home folk would profit."[26]

The desire to promote high scientific standards for popularization was an essential driving force in the coalition. At the turn of the century, biologists had resoundingly condemned certain popular writers and labeled them "nature fakers" for distorting the roles of science and sentiment in their nature and wildlife stories and for ascribing to animals unrealistic actions or behavior. Rampant industrial development, expanding population, encroachment of suburbs on previously rural areas, proposals to set aside national parks for leisure, the rise of scientific understanding of the natural world, and the assertion of science's rational primacy over religion and spirituality had all stimulated cultural debate about the appropriate relationship "between people and nature," observes historian Ralph H. Lutts.[27] Criticism had rattled through the halls of natural history departments such as Clark's. As a result, the biologist argued against including certain nature study or conservation groups in the radio alliance on the grounds that they were more concerned with "sentimentality" than scientific accuracy. "The maintenance of the prestige of science before the general public, which means the supervision by competent men of the channels of popularization, is one of the most important problems today," Clark wrote to John Phillips. Clark assured the other potential collaborators that, even though Burgess was not trained as a scientist, he shared their concern for accuracy. Moreover, "because of his great popularity," Burgess "can do more than all of us together to keep popular science within safe bounds."[28]

The Boston natural history talks also offered a way to promote biology. The physicists, Clark explained, seemed to be attracting all the publicity in the 1920s; it was time for biologists to "come in with a series of 'talking points' that will present biology in a new light and appeal to the popular...."[29] "Just now the physicists, chemists and astronomers have the public ear as a result of their ingenuity in presenting their case in the shape of new and constantly changing aspects," Clark told Wigglesworth.[30] Surely the biologists and those interested in natural history could do as well in promoting their own achievements to the public.

CONSERVATION AND SCIENCE

In January 1926 Burgess celebrated the first anniversary of the Radio Nature League, delighted at "spreading the gospel" of conservation and dreaming of opportunities to do more.[31] Throughout his career, he frequently lent his voice to other conservation campaigns (for example, appearing on a program about "The Effect of Forest Fires on Wild Life" during observance of American Forest Week) but he also initiated several large-scale efforts himself. His approach was clever and attuned to human psychology: Conservationists should "lead rather than force" the public to protect wildlife, should attempt to change attitudes rather than relying only on enactment of laws and regulations. Most people are, he believed, "thoughtless rather than malicious" in their treatment of nature, so he sought first to "stimulate their interest and arouse within them the desire to save the wild life" rather than scolding them for prior bad acts.[32] Radio provided a convenient venue to test this approach. When Burgess started a campaign to convince New Englanders to refrain from picking the blossoms of a threatened wildflower, he simply asked listeners to send in pledges. "It would do your heart good to read the pledges coming in never again to pick fringed gentians," he wrote to Clark—and then added by hand at the bottom of the letter "Darned if I don't believe it is all worth while!"[33]

One of his more successful projects involved encouraging listeners to destroy the egg masses of tent caterpillars that were defoliating sugar maples and various ornamental trees. New England was in the midst of an extensive regional outbreak of the pest. The moth lays several hundred eggs at a time and attaches them, with a gluelike substance, around a tree branch. It is easy, albeit time-consuming, to disrupt the caterpillar's life cycle by destroying these masses. In March 1926 Burgess inaugurated "Tent Caterpillar Week" and offered book prizes to the schools, Boy and Girl Scout troops, and individuals making "the best record in the gathering and destruction of Tent Caterpillar egg masses." "Promptly on my announcement of this contest," Burgess told Clark, "700 scouts were enrolled from one of the eastern counties of New York and are hard at it"—"I believe it will accomplish a great deal for the betterment of our roadside beauty."[34] Burgess ran another campaign the next year, for three weeks during early spring, which resulted in the destruction of well over a million egg masses.[35] By 1928 listeners who wanted to compete in the contest were complaining that they could not find as many egg masses as in previous years ("In other words, they cleaned them out a year ago and this year shows the result"), but a newspaper that had offered a $100 prize in previous years was so pleased with public reaction that the amount was raised to $150 in March 1929.[36] Despite the work involved in tabulating the responses, Burgess continued the annual campaigns because participants could see the results ("Tent caterpillars are becoming very scarce along the roadsides in certain sections as a result of these campaigns").[37]

Burgess also did not shy away from using his broadcasts for political purposes. Conservationists were attempting during the 1920s to persuade the U.S. Department of Interior to restore an Oregon lake originally drained to assist Western agricultural interests. By 1922 only a few hundred acres remained of a fertile wetland area that once spread over eighty thousand acres and had been both a rich breeding ground for resident birds and wildlife and an important shelter for migratory wildfowl along the Pacific flyway. Local citizens, conservationists, and scientists were campaigning to reflood some of the area, so Burgess invited conservation advocate and wildlife photographer William L. Finley to describe on the air the plight of Lower Klamath Lake: "He is fighting to get the water turned back into the lake bed to cover some twenty thousand acres of worthless land."[38] After Finley told the story, Burgess asked League members "interested enough to want to save our bird life, no matter where located, to send in their names for a petition to the Secretary of Interior, asking him to open the gates at Lower Klamath." Burgess interpreted the public reaction as proof of what radio could accomplish. "The names are pouring in with the most interesting letters," he reported to Clark, who in turn publicized the Klamath broadcast to other scientists at the Smithsonian and elsewhere in Washington as an example of how to advance the cause of conservation.[39] Two months later, Burgess sent an update: "You might like to know that the names still come in for the Klamath Lake petition and the other day I received from one woman one thousand and one names. Yes, we can do a whole lot on the radio."[40] Although the Department of Interior's Bureau of Reclamation resisted reflooding Lower Klamath Lake, the nationwide publicity campaign eventually produced a compromise; in 1928 President Calvin Coolidge withdrew some of the lands from the reclamation project and established the Upper Klamath Wildlife Refuge (later named the Tule Lake National Wildlife Refuge) to be administered by the U.S. Department of Agriculture's Bureau of Biological Survey.

Using radio for popular education or political action was, for Clark, not sufficient. He was convinced that the medium had significant potential for "gathering as well as giving out scientific information." Determined to try "to find some means of using the radio for scientific ends," he persuaded Burgess to cooperate.[41] In the summer of 1925, for example, there was an unusual flight of three species of white herons in Canada and the northeastern United States. When Burgess asked listeners to submit descriptions of sightings, he received reports of nearly a hundred of the birds, and many of the reports were so accurate that researchers could tell whether a listener had spotted a snowy egret, American egret, or immature little blue heron, all of which were then in white plumage. In 1927, also in coordination with researchers, Burgess asked listeners to report sightings of snowy owls (*Strix nyctea*) along the U.S. East Coast.[42]

The most elaborate data gathering effort centered on attempts to understand the periodic fluctuations of species such as the ruffed grouse. The Massachusetts Fish and Game Protective Association (MFGPA) was already conducting research on the grouse when Clark persuaded his friend John C. Phillips, a sportsman and MFGPA committee chairman, to appear on the Burgess show during the fall hunting season. Phillips asked listeners to send both diseased and healthy birds (either shot or accidentally killed) to scientists Alfred O. Gross at Bowdoin College and Arthur A. Allen at Cornell University, who were analyzing the birds' crop contents for the presence of disease, parasites, or dietary indicators that might explain their dramatic population swings. Over seven hundred specimens were submitted for autopsy, along with additional useful information about where they were found. Clark and many other scientists regarded this project as additional proof that radio could be a "valuable medium for collecting information of scientific interest and political importance," and a useful "adjunct to scientific investigation."[43] He later attempted to persuade anthropologist Alfred V. Kidder to make a similar request for information relating to "Indian sites" and "shell heaps" and thereby "make the radio serve the ends of anthropology in New England."[44]

Like Clark, Burgess was convinced that adults could be attracted to natural history subjects through radio popularization. In contrast to the elitists, though, he believed that the fault for communication failures resided more often in the sources than in the recipients. When people seemed disinterested in natural history, he believed the communicator might be to blame: "Even the educated man is more or less of a child when it comes to liking his information well sugared and in exceedingly simple form."[45]

Despite his popularity as an author, his solid reputation, and sporadic attempts to develop a national series, Burgess never hosted a series broadcast beyond New England. Network executives in New York apparently pegged him as "Burgess of the Bedtime Stories," incorrectly perceiving him as only a children's host rather than as the creator of a conservation show with a substantial and loyal adult audience. Burgess continued to arrange occasional hookups between the Boston stations and those in Washington, but the radio industry was changing. His experience is characteristic of a number of well-done regional radio series on science and conservation. By the late 1920s, without commercial sponsors willing to pay network rates, these popularizers could not gain the opportunity to compete for bigger audiences on national platforms; moreover, even at the local level, scheduling was increasingly affected by network rather than regional executives. Scientists began to be pushed to the margins or off the air altogether.

ZOO TIME

The uncertainty of success in broadcasting did not deter others from trying. In spring 1925, entomologist William M. Mann was appointed superintendent of the Smithsonian's National Zoological Park. At thirty nine, Mann was a vigorous, gregarious person, fascinating to be around, with an infectious passion for life. Although Clark and Mann shared Harvard connections (Mann had received his Sc.D. from the Bussey Institution at Harvard), they had become friends when Mann began working at the U.S. Department of Agriculture's Bureau of Entomology in Washington in 1916. As soon as Mann was named head of the zoo, the two men began to plan a second Smithsonian radio series centered on animals and modeled on nature programs being offered around the country, such as by the Bronx Zoo.

Clark promised that this program would be "the natural outlet for nature study talks in Washington."[46] He eagerly took on the role of arranger and impresario, handling all arrangements with the station and special guests, and leaving it to Mann to supply the colorful anecdotes and engaging personality. "The new series of radio talks . . . will come out under [Mann's] name, though the idea is mine and I shall make all the arrangements for it," Clark wrote to Burgess. "Mann's personal charm and great popularity insure the success of such a venture."[47]

"Radio Nature Talks from the National Zoological Park," a fifteen-minute presentation preceded by current news from the zoo, premiered at 8:15 p.m. on Saturday, October 3, 1925. Following Everett F. Haycraft's "Bible Talk" and preceding concerts at 8:30 and 9:00 by Sherry's Orchestra and the Dexter Male Chorus, the first program ("Introduction to the Zoo") was an immediate success. Other programs that season included Mann describing "Collecting Wild Animals in South America" and "Snakes of the District of Columbia" and Smithsonian scientist Doris M. Cochran discussing "Giant Tortoises" and "Alligators and Crocodiles."

This series also became a family affair for Clark. On October 31, 1925, his twelve-year-old son Hugh appeared in a scripted conversation with Mann ("What a small boy wants to know about the Zoo"). Hugh Clark's questions centered on giant anteaters, hyenas, and the cry of the tapir, all topics chosen to allow Mann to describe creatures he had encountered on expeditions or "to serve as a vehicle for personal anecdotes."[48] As they prepared the final script, Clark encouraged Mann to "run in as much of the personal element as you can throughout," and explained that "this is the sort of thing the public likes."[49] Clark's instincts again proved correct. Over a hundred newspapers around the United States mentioned the broadcast.[50]

In 1926 Mann led the Smithsonian-Chrysler expedition to Tanganyika to collect animals for the zoo. The Smithsonian hired a New York public relations firm to market expedition photographs and to arrange exclusive press coverage. Although the promotions manager declared that they intended "to keep the publicity on a high plane, dignified, in keeping with the importance of the Smithsonian Institution, scientifically correct and accurate in every detail," Pathé News was not discouraged from placing "six men dressed as lions and monkeys" on the dock when Mann sailed in March from New York ("they added a bit of color to the picture").[51]

Throughout the summer, Clark worked closely with the public relations firm, approving copy on Mann's behalf and sometimes suggesting topics for stories:

> Write up some of the dangerous insects in Africa—the tsetse fly, the Congo floor maggot. . . . Write up the queer fishes, the electric cat fish, the fresh water flying fish, the fish that in dry season makes cocoons in the mud. . . . The giant frogs . . . the pythons, tree cobras . . . offer points of interest.[52]

Clark continued the zoo programs in Mann's absence by inviting guest speakers and, in the first two to three minutes of each broadcast, reading from the scientist's colorful letters about the sea voyage to Africa or the expedition's attempt to track a herd of gnu or trap a young rhinoceros. Once again, Clark's instincts were good. By July, Clark explained (good-naturedly) that he had been "relegated to the position of manager for the prima donna," and he wrote to Mann that the "letters are making a great hit here. . . . In the opinion of the elevator and check boys at WRC . . . you are the world's leading hero."[53] In January 1927 Clark arranged for his daughter Sarah to appear in a dialogue with Mann while the zoo director recalled his experiences on the Tanganyika expedition.

Within a year, with the help of Clark, the radio show, and the Smithsonian's public relations firm, Mann had "metamorphosed from an entomologist entirely unknown except to his coworkers into the most widely advertised zoo director in the country."[54] The zoo series, sometimes focusing on Mann and other times on descriptions of news or announcements of events at the park, although less on science or conservation, continued in various formats for many years.

SHAPLEY AND THE STARS

Another of Clark's important contacts during that time was with astronomer Harlow Shapley, whose brief series of broadcasts reflected the conservative

FIGURE 3. Harlow Shapley, director of the Observatory at Harvard College. The astronomer, shown seated at his revolving desk during the 1920s, engaged in popularization efforts throughout his life, including via radio. Courtesy of Smithsonian Institution Archives.

approach adopted by most scientists who dabbled in radio during the 1920s. The forty-year-old Shapley was already on the way in 1925 to becoming one of America's most visible astronomers. At Princeton he had worked with Henry Norris Russell, and he had been director of the observatory at Harvard for only four years. Throughout his career, Shapley participated enthusiastically in popularization, from writing magazine articles to spearheading a shortwave radio station focused on educational broadcasting. In 1920 he had begun what would be a long association with Science Service, an institution that had just been created to promote popularization (see chapters 3 and 4).[55]

Shapley apparently caught the "radio bug" soon after conversations with Clark. In September 1925 Burgess reported that discussions were "bearing fruit" and that the astronomer had suggested developing radio talks based on the observatory's work. "I understand that the program manager will turn him over to me to work these talks in on my programs," Burgess wrote. "It will be bully if this comes through."[56] Shapley, however, decided to arrange the programs himself. They were broadcast from November 3, 1925, to January 19, 1926, on WEEI and coordinated with explanatory newspaper columns the following mornings in the *Boston Herald*.

Although brief, the Harvard series is especially notable because four of the twenty-two talks were delivered by women scientists, Cecilia H. Payne ("The Stuff Stars Are Made Of" and "Stellar Evolution") and Annie Jump Cannon ("Classifying the Stars" and "New Stars and Variables"). Almost half of the talks were repeated (reread) on Clark's Washington broadcasts.

Although the Harvard series aimed to educate about astronomy, Shapley could not resist the temptation to titillate his audience in the final talk with one of radio's favorite subjects—"Life in Other Worlds?" "There seems to be an inborn fear of loneliness on the part of the ordinary individual when he contemplates the possibility of other worlds," Shapley declared:

> We are disappointed if the cold-blooded scientist assures us that Man cannot exist on Mars. . . . We do resent a restraint on our imagination. We should like to believe in marvelous men on Venus.[57]

It was a speculative talk. Dreams and imaginings intertwined with Shapley's cautious assessments of where life could not survive and where, just possibly, a living complex, a planet hospitable to nonterrestrial life might exist. This clever theme, one bound to have excited listeners, echoed over the airwaves repeatedly in the following decades.

Shapley arranged one more show on which he answered listeners' questions, with the responses printed in the following Sunday's *Boston Herald*.[58] As he recounted to Clark, they "received, of course, too many questions, but that gave me an opportunity to select, and all went merrily."[59] He also boasted that his programs' success had had the unanticipated but positive consequence of inducing the "rather stiff-necked" Boston Symphony Orchestra "to go on the air last Saturday night, beginning a series of broadcasts, because the Harvard broadcasts had lent a dignity of a sort to radio."[60]

Although Shapley continued as a dependable radio guest for many decades, this was apparently his only substantial effort to produce a series himself. To arrange such broadcasts borrowed considerable time from a scientist's research. It also required a strong commitment from the organization. Shapley later told Clark that he had met with the university's president, A. Lawrence Lowell, to discuss "what the attitude of Harvard toward broadcasting may be in the future" and that he had recommended to Lowell that the university be purposeful in its involvement in radio, probably through the Division of University Extension. But, Shapley added, whatever Harvard attempted would have to have "purpose and an element of sequence" because "the single, isolated, and unrelated radio talk was pretty much a shot in the dark."[61] As Shapley recognized, radio popularization required resources far different from those required for a university press or magazine.

MOTIVATIONS

For universities and institutions like Harvard or the Smithsonian, involvement in radio throughout the 1920s and 1930s was, in fact, more often sporadic than purposive, a pattern of institutional behavior that continues today. Opportunities to engage in broadcasting would arise, but even when they were seized, the organizations' commitment to popularization was rarely deep or long-term. Most often, the motivation for involvement derived not from any sense of obligation to engage in civic education but from an (untested) assumption that the "good publicity" provided by a radio presence would increase public support and therefore attract funding. The Smithsonian, for example, characterized radio from the outset as both a way to accomplish its chartered mission, to "diffuse knowledge," and an adjunct to its fundraising.[62] The institution first attempted to project a radio "presence" while also launching a major endowment campaign, and at a time when scientists around the country were seeking political support for national research endowments.[63]

For individual scientists, the motivations were more complex, especially because time spent on popularization is time spent away from research. Whatever the field, researchers spend years learning their craft, building professional reputations, and gathering and analyzing data. They study stars, starfish, auks, or grouse because they are passionately interested in them, content to spend every waking hour thinking about science at the desk, in the lab, in the field, in the bathtub. Why, then, break concentration and momentum to go on the radio? Each of these scientists tended to view a broadcast's purpose through the lens of his or her own particular discipline or interests— it would improve understanding of biology or astronomy, it would advance the cause of conservation.

For some, personal ambition also played a significant role. The involvement of Clark, for example, whose political attitudes were far from populist, can seem especially puzzling until we look more closely at his actions and statements in February 1927. Within one day of the sudden death of Smithsonian secretary Charles D. Walcott, Clark was actively campaigning to be Walcott's replacement. In that effort he sought the support of previous radio guests and of the science journalists he had assisted.[64] For Clark, the radio work had been a way to advance his career generally and to further his ambition to become head of the Smithsonian (a goal ultimately thwarted by the appointment of astronomer C. G. Abbot).

Burgess appears to have had more modest aspirations, but he too was not immune to the value of publicity. Radio helped book sales. "I am the direct product of advertising," he wrote:

> There is no use trying to sail under false colors. It is the constant printing of my name day in and day out, year in and year out, which has given me such standing as I have in the public mind. Even Lydia Pinkham became a household name. Publicity did it.[65]

Despite differing motivations and differing outcomes, those first radio ventures of the 1920s signaled an important change in scientists' acceptance of popularization through media formats other than print or formal lecture. Individuals like Burgess, Clark, Mann, and Shapley succeeded not only because of foresight, ingenuity, and commitment to public outreach, or because of their ability to persuade other scientists to become involved. Their efforts were also facilitated by concurrent strengthening of science popularization in other venues and the founding of an organization whose sole purpose was to promote the cause of popularizing science, including via new formats like radio.

CHAPTER THREE

———— ✻ ————

Syndicating Science

Collisions of the atoms displaced automobile and railroad collisions; slaying of bacteria and undesirable insects completely overshadowed similar "activities" among humankind; pictures of scientists ornamented the pages hitherto decorated by pictures of statesmen and criminals. Believe me: the scientist had his "day" in the way of publicity this time.   HARRY L. SMITHTON, 1924[1]

WHILE individuals like Austin Clark, Thornton Burgess, and Harlow Shapley were experimenting with their broadcasts, two organizations in Washington were joining forces to create another new science series. In the years following World War I, a newfound interest in public relations had rippled through the scientific community. The partnership between the National Research Council (NRC) and Science Service signaled to other scientists that radio could be an appropriate and dignified platform for popularization and for achieving positive publicity for all science. Although the NRC eventually relinquished its direct role in the series, Science Service continued to be involved in radio production for the next forty years, essentially functioning as the scientific establishment's sanctioned surrogate on the airwaves.

To understand Science Service's commitment to broadcasting requires first understanding the organization's founders and essential mission and the important role it played in sculpting American scientists' attitudes toward popularization during the 1920s and 1930s. The organization's creation was rooted in a noble idea: "to reach the widest possible audience with the largest amount of scientific information" and to do so with accuracy and diligent attention to the audience's needs and interests.[2] Its success came from an

45

ability to exploit techniques of American commerce and advertising to feed the expanding market for science, primarily through the print media but also by acting nimbly to experiment with other modes of communication like radio. That success not only demonstrated to editors and publishers that science content could be made interesting to millions of their readers, but also gave popularization increased legitimacy as an activity deserving of every scientist's time and attention.

MUTUAL CONCERNS

The idea for a science news service emerged from the friendship of two extraordinary men—scientist William Emerson Ritter and newspaper publisher Edward Willis Scripps—who shared an interest in marine biology and the social sciences. "From the beginning," Mary Bailey Ritter recollected later, her husband's

> great interest in the study of man from the standpoint of the highest type of animal . . . particularly interested Mr. Scripps. For he [Scripps] was studying man from the [standpoint of the] hundreds of men employed on his newspapers.[3]

Their first interactions occurred when Scripps became a patron of Ritter's research. Scripps had built an impressive publishing empire, eventually establishing or purchasing twenty-one newspapers and creating the Newspaper Enterprise Association (a news and photo feature service) and the United Press Association (a national wire service).[4] For relaxation, the millionaire began spending increasingly more time at Miramar Ranch, an estate fifteen miles northeast of San Diego, California. Every summer, Ritter, head of the department of zoology at the University of California at Berkeley, took his students for research along the southern California coast near the ranch. In 1902 Ritter discussed the idea of a new marine biology laboratory with Fred Baker, a prominent San Diego physician fascinated with Mollusca research.[5] Baker persuaded his friend and poker partner Scripps to visit Ritter's temporary summer research lab. That was, Mary Ritter recalled, the beginning of a twenty-year friendship between the two middle-aged men.

As the search began for possible locations for a permanent laboratory, Scripps suggested that he buy a large tract of undeveloped shore land and, with his sister Ellen Browning Scripps, who lived nearby, endow a laboratory in memory of their brother George Scripps. Other prominent San Diegans joined in the campaign to establish the marine station, Scripps donated a schooner for ocean research, and Ritter became director of the "Scripps

Institution for Biological Research" (later named the Scripps Institution of Oceanography).

These two inquisitive, intelligent men must have had extraordinary conversations as war raged in Europe and then peacetime life, culture, and politics careened into the twentieth century. What would America's citizens need to live, to work, to thrive? Both Scripps and Ritter believed strongly in education, but they were also convinced that the traditional school curriculum, while important, would never be sufficient. The working classes, and the waves of immigrants, needed information to equip them for unfolding challenges, for technologies and social change barely visible on the horizon. The "master problem," Ritter wrote, was "how shall the great rank and file of the population be assured that minimum of natural knowledge without which prosperity and progress and happiness in the modern world are impossible?"[6] Such "natural knowledge" would not spring from the agrarian experiences of Ritter's childhood or from the skills honed in factories or Scripps's pressrooms. Nor, he added, would the answer emerge from a science "reduced . . . to the narrow, inhumane bounds of industrial technology." The populace, Ritter argued, must be educated in—or, at minimum, made aware of—both "traditional" science ("indisputable objective knowledge of whatever kind or source") and multidisciplinary, interdisciplinary science (what he called "a structural foundation to build on," a science for living as well as working). Such an approach to mass education fit Scripps's long-standing emphasis on newspapers as advocates for the interests of the working classes, for he believed that the press should engage in aggressive civic reform without indulging in "yellow journalism" simply to sell newspapers.[7]

Sometime in 1919, Scripps and Ritter began to discuss the formation of an organization to implement this vision. This agency for "life-continuation education" would demonstrate how to discuss science both interestingly and accurately. Scripps spoke of forming a "Gideon's Band" of scientists and scientifically trained journalists to engage the masses while promoting "the value of research and the usefulness of science," under the watchful eye of a governing board composed of establishment scientists and representatives of the newspaper industry.[8]

For two months during the winter of 1919–1920, Ritter visited several hundred research installations around the United States, soliciting comments on Scripps's idea from more than three hundred scientists and journalists, and building support for a project to use newspapers "for promoting a wider knowledge of the results of scientific investigation and a more general appreciation of the methods and spirit of science."[9] Ritter later claimed that only "about five percent" of the scientists were "positively hostile" to the idea, while at least half expressed a willingness to help the enterprise. The rest were willing to suspend judgment until operations were under way. Almost all agreed on one aspect: what was needed was not an umbrella organization run by one of the existing associations. This task called for something new, inclusive, innovative, and designed to fit the proposed public service goals.

THE CONTEXT

For the scientific establishment, the notion of an entity devoted exclusively to popularization was neither repellent nor automatically welcome. During the first decades of the twentieth century, the scientific community's indifference (and occasional hostility) to popularization had begun to evolve into a begrudging acceptance grounded in practicality. Scientists recognized the potential political and economic benefits of cooperating in public dissemination of their ideas; with active involvement, they might be better able to monitor accuracy, certify authenticity, and claim credit, and such attention would surely encourage increased political investment in science. There was also an expanding potential audience for news and information about science. Consumer technologies, relativity, advances in chemistry, and archeological expeditions all were contributing to a sense of intellectual excitement at the same time that a voracious media market—newspaper chains and large-circulation general magazines—was searching for interesting content to repackage for readers.

Science had emerged from the Great War with an enhanced but somewhat ambiguous image, primarily because of the publicity that attended the development of chemical weapons. Science's benefits were undeniable. Chemists and physicists had contributed to military and defense preparedness; public health and medicine were extending human life; genetics, entomology, and soil chemistry were improving agriculture, killing pests, and increasing crop yields; and the technological products of physics and engineering (like wireless sets and phonographs) made the Jazz Age sparkle. Relatively few American researchers received government funding, however, unless their work had immediate application to public health, agriculture, or weaponry. Scientific leaders in the United States saw a direct connection between money and public image.

University of Colorado entomologist T. D. A. Cockerell expressed the frustration of many of his colleagues when he asked in 1920: "Why Does Our Public Fail to Support Science?"[10] Although his own state legislature had

declared that it "believed in education" and enthusiastically supported college athletics, the politicians expressed little comprehension of why research
was an equally essential function in a university; some legislators had even
suggested that research "can wait." Such dismissive attitudes, Cockerell argued, were related directly to the public's image of science. Cloistered away
in their laboratories, shielded from public observation, researchers could
easily be perceived as harmless eccentrics engaged in woolgathering rather
than significant contributors to a "national" activity directed at the greater
good. "Scientific research is not a thing isolated," Cockerell explained; "it is
part of the necessary work of the world, and when once that is understood,
it will take its place along with our other normal activities."

> With a definite program, sufficient means for publication, and dignified arrange-
> ments for publicity . . . there would be no difficulty in obtaining the support of a
> majority of the citizens. . . . But it is not sufficient to preach the virtues of science
> in general. People must be shown in detail, in a thousand ways suited to their
> particular needs and interests.[11]

Today, initiating a public relations campaign to change attitudes and
improve a profession's reputation seems standard practice. In the 1920s it represented a radical approach for academic researchers, but it conformed to
the times. Advertising was aggressively marketing all sorts of dreams, guar-
anteeing satisfaction, telling endless "fables of abundance."[12] Americans were
being sold promises of self-improvement, social progress, and spirituality
along with their toothpaste, furniture, and Fords. Why not "sell" science as
well?

Cockerell's word "dignified" hints at the type of forum that scientists considered appropriate. Print had always been the preferred choice, followed
by public lectures and museum exhibitions. The large-scale publicity campaigns of the National Research Council, American Association for the
Advancement of Science (AAAS), and American Chemical Society (ACS) involved books and magazines. The Chemical Foundation was subsidizing a
project to sell six popular books on chemistry, at a nominal price and
promoted through full-page newspaper advertisements promising "The
Progress—The Romance—The Necessity of Chemistry!" The foundation donated twenty thousand sets to schools and public libraries.[13] What scientists
like Cockerell envisioned was not the rowdy sensationalism of the tabloid
press but something more suitable for a library—a controlled assemblage of
authenticated descriptions of accomplishments and insights.

Scientists alone could ensure the necessary accuracy and authenticity, yet most tended to regard popularization as an avocation peripheral to scientific life. How, then, would the scientific community accomplish its goal of selling itself? One approach was to exploit professional journalists to do the front work, that is, for the scientific associations to establish press offices to facilitate news coverage. The American Chemical Society created its permanent news service in 1919 and, in less than a decade, was spending about $8,800 a year on press relations activities, feeding news copy and information to reporters.[14] AAAS chose not to consolidate press operations in a single office; instead, the association sought to influence content by cultivating good relations with journalists, making them feel comfortable at meetings, and speeding the flow of information from its meetings by arranging special interviews and releasing advance copies of talks.[15] The AAAS executive committee stated unequivocally that "the present attitude of the newspapers toward science should be encouraged as far as possible and that newspaper men at our meetings should be given every possible aid."[16]

By 1920 newspaper attention to science had grown, the number of science articles in general magazines was increasing, and publications like *Popular Science Monthly*, *National Geographic*, and *Scientific American* had healthy circulations. The number of newspaper articles on science doubled between 1920 and 1925.[17] The proportion of science articles that were the lead or cover articles in general magazines also rose in the 1920s.[18] And during the years after the war, a group of leading scientists spearheaded a campaign to establish their own popular science magazine to publicize the value of research.[19] Between 1915 and 1919, members of an AAAS committee held discussions with potential publishers, explored a proposed cooperative arrangement with ACS, and fended off the complaints of James McKeen Cattell (publisher of *Popular Science Monthly*). Ultimately, disciplinary jealousies, lack of funding, and the lingering concern that popularization activities would distract researchers from their work kept the magazine from advancing beyond the planning stage.[20]

Scripps and Ritter launched their proposal at an auspicious moment: scientists were looking more favorably upon popularization but were not quite willing to fund efforts themselves. The Scripps-Ritter approach emphasized authoritative and accurate as well as intelligible content; it stressed cooperation among the disciplines; and it left scientists feeling in control of their own intellectual products. Had the idea come from scientists alone, journalists might have regarded it with suspicion and perceived it as self-serving. Had it come from a news organization (or just from Scripps), then scientists would have been leery of the potential for sensationalism and uneasy at the lack of control. Together, though, Scripps and Ritter were a powerful combination—a rich, successful, smart newspaper executive and a respected and visionary scientist offered a grand vision of a civic society informed about science and technology. Their rhetoric in published essays and private letters was lofty, their instincts pragmatic, and their ideas often uncannily prescient about what the nation's citizens would need as the century unfolded.

THE VISION

Scripps had made a fortune because he understood the business of producing and selling news. While rival publishers Joseph Pulitzer and William Randolph Hearst experimented with news and entertainment content, Scripps "experimented with the news business," developing approaches to management, performance, circulation, revenue, and workforce that were all later adopted by his competitors.[21] This philosophy inevitably influenced the direction of the proposed science popularization entity. The organization would not propagandize for science but would instead market the idea of useful information about science and seek to persuade the press to pay more attention to science. Whether characterized as a "Society for the Dissemination of Science" or a "Science News Service" (the first two proposed names), the goal would be "to CAUSE the dissemination of science, which does not necessarily mean direct action."[22] The group would leverage news coverage by developing materials that others could use and by educating editors about the importance of science news. "In the business world it is common enough to speak of the 'creating of a demand,'" Scripps explained to the first director. "Anything that you could do in the way of attracting the attention of journalists to the subject of science will naturally create a demand for your product—and what is even more desirable, will create a demand by editors for scientific matter generally."[23] Although neither Scripps nor Ritter initially described the organization as concerned with any format other than print, they emphasized that the activities should be centered on disseminating the fruits of scientific research via "any available means . . . except formal school work."[24]

Scripps provided a comfortable but only partial endowment, intending that Science Service would invest his annual payments and use the resulting income to subsidize projects; the accumulating endowment would cushion against changing economic conditions or failed experiments.[25] Even though

Science Service would not be seeking to make a profit, it should always be run like a *business*, and it was set up as a not-for-profit corporation. The group must also charge a fair price for most services, Scripps believed, because something provided for free would be "valued" less by the recipient. This notion that Science Service should be self-sustaining initially attracted skepticism from the scientists—probably because they did not perceive news about science as a commodity for which there might be a market. Ritter, however, supported the idea, writing to Scripps:

> Unquestionably there are aspects of science that appeal strongly to popular interest. There is much that is curiosity-satisfying, much that is practically useful, much that is dramatic; and were Science Service to "play up" these aspects to the extent that it might it could soon reach a self-supporting basis, and could go on and largely increase its funds.[26]

Economic self-sufficiency in fact helped Science Service maintain a certain independence from the scientific community during its early years, although that relationship was tested severely during the early 1930s. Scripps believed adamantly that the organization must not be perceived as—or, indeed, be—a publicity machine for science. In numerous letters and documents, he argued against "propaganda." Science Service must not support partisan causes, including that of science itself; it must be a reliable source of facts, a translator of complex ideas.

Although skeptical of untrammeled charity ("The man who casts his bread upon the water in order that it may be returned to him manyfold is the worst kind of usurer and profiteer"), Scripps expressed satisfaction with Science Service:

> It made me feel that it is quite possible that I have been fortunate in stumbling on a scheme for which I could disburse some money to great public advantage. The very words, charity and philanthropy, cause me to have a feeling of nausea on account of my own experience and observation having proved to me that "charity" and "philanthropy" are indulged in more generally by those who are vulgar, ostentatious and ignorant—men and women who do infinitely more harm than good. . . . What I felt after reading your report was that I had found a way of giving away some money without the least possibility of doing anybody any harm.[27]

Ritter's views ran along similar lines. He also expressed strong opinions about who should assume the main responsibility for popularization. In 1920 Charles B. Davenport had asserted to Ritter that "the right way to bring the results of scientific research to the general public" was through skilled science writers.[28] Ritter responded that, yes, writers "who are highly intelligent and sympathetic ... though not necessarily trained in science" could visit laboratories, observe scientists at work, read the investigators' technical reports, and engage in the work of translating science to the public, but trained cadres of writers would not be sufficient. It was not just a matter of who wrote the popular accounts. Scientists must also change their own attitudes toward popularization.

Science, Ritter explained to Davenport, must determine how to take "its natural place in the total frame and fabric of civilization," how to get itself "into right relations with the other major factors of civilization." The effort to popularize science must proceed with "considerable sympathetic attention to the problem of the deeper social relation of science." Philosophy, logic, and physics have been struggling with the notion of "relation as a phenomenon of nature," but now science must understand its "relation" to society and scientists must "see more clearly and assume more fully" their "responsibilities to the public at large," especially in "working the spirit and substance of science into the community mind." Scripps had convinced Ritter that not all researchers should (or would) "fall forthwith to writing 'stories' for the daily papers about their researches." Nevertheless, scientists' cooperation was essential. Researchers must be convinced of the usefulness and importance of Science Service, must see how its work could help them, and must become engaged in the work. "The scientific men themselves have an extremely fundamental obligation here which they cannot possibly shunt off to anybody else," including to the usual popularizers. Attempting to shift that obligation to professional journalists alone might save them time (scientists prefer to be in the laboratory rather on the public platform), but it would cost them down the road. For Ritter, "The obligation of science ... to discover its own meaning for human life is not one whit less pressing upon it than is its obligation to discover the truths of nature."[29]

Science Service's governance structure incorporated that vision of shared responsibility. Scientists (serving on a board of trustees) would provide a continual check on accuracy, and would smooth access to the latest and most reliable research. Scientists would provide endorsement, sympathy, and support, and would occasionally become more actively engaged. Journalists—prestigious editors and publishers—would advise on the news business and would assist the syndication operation through feedback on the salability and appeal of the organization's products, whatever the medium and format. The audience would then reap the benefits of their collaboration.

FIGURE 4. Engineer and inventor Hiram Percy Maxim (left), founder of the American Radio Relay League, with Edwin Emery Slosson, director of Science Service. Slosson attended scientific gatherings around the world in the hunt for sources, potential authors, and radio guests to enrich his organization's work. Courtesy of Smithsonian Institution Archives.

GEARING UP AND ACQUIRING CONTENT

During the summer of 1920, the planning committee met with Scripps at Miramar Ranch to consider proposed candidates to head the organization.[30] Soon afterward, Ritter began negotiating with a well-known chemist and popularizer, Edwin Emery Slosson. That December the final details of Science Service were forged during another organizational meeting of scientists and prominent journalists in Chicago. Slosson arrived in Washington in January 1921 to set up an office and begin work.

Slosson turned out to be an inspired choice. Within a year of graduating from the University of Kansas, he had started teaching chemistry at the University of Wyoming and also begun part-time graduate study at the University of Chicago. Slosson's popular articles on chemistry soon attracted the attention of publisher Hamilton Holt, who in 1903 invited the chemist to move to New York City and become literary editor of Holt's magazine, the *Independent*. By 1921, at age fifty-five, Slosson had acquired a formidable reputation as a popularizer. He was author of *Easy Lessons in Einstein* and other well-received books. The Chemical Foundation had just distributed sixty thousand copies of Slosson's *Creative Chemistry* as part of its publicity campaign.[31] Writing, not research, had become his passion. As he confessed to a fellow chemist who had just entered law school:

> I too like you am classed as a "renegade from natural science" since I have never done any research work in chemistry after having taken my doctorate at the University of Chicago in that science. But I have like you retained my interest in science and have done what I could to spread a knowledge of scientific achievements among the reading public.[32]

Formal announcement of the establishment of Science Service ran in the April 8, 1921, issue of *Science*. In the issue's lead article, Slosson described "A New Agency for the Popularization of Science" intended to "bridge" the gap between the minority of the population that "habitually read the scientific journals" and "a majority that never touch even the most popular of them." Echoing the rhetoric and politics of Scripps and Ritter, Slosson declared that the success of democratic societies depends on their citizens' abilities "to distinguish between real science and fake, between the genuine expert and the pretender." By establishing standards for reporting on science and by helping to educate newspapers about science, he wrote, Science Service would "act as a sort of liaison officer between scientific circles and the outside world."

Slosson had hit the ground running, writing letters to potential supporters and contributors even before he had left New York City. He embarked on an ambitious campaign to build a news operation without hiring full-time professional journalists to do all the reporting. This was perhaps a strange way to set up a *news* syndicate, but it worked. Through the years, the Science Service staff became extraordinarily skilled at taking scientific information, news, and even ideas produced by others and repackaging them for sale. They filtered out most of the cranks, cross-checked for accuracy, added definitions, and translated the technical language. To a beat reporter accustomed to reconstructing the sequence of an urban event and to interpreting and confirming facts by locating witnesses, the Science Service approach may not have seemed like authentic journalism. For science journalism—a profession in embryo—the process made sense. "News" about science unrolled deliberately, at the pace of the research itself. Scientific "events" like eclipses were anticipated, planned for, recorded carefully, and then interpreted over months or years. The notion of "breaking news" or "hot news" in science usually did not make sense.

Nevertheless, the news articles the organization marketed could not be stale. To acquire privileged information in a timely way, its staff needed the cooperation of the entire scientific community. Slosson encouraged young men and women in the various sciences "who have literary inclinations" to submit popular articles (which Slosson then marketed to periodicals and newspapers, acting as an agent and raking off a percentage of any royalty), to send their papers and talks in advance of scientific meetings, or even to describe others' research in short news items (for which Science Service paid them a small fee upon acceptance).[33] Establishment of an active network of these stringers—writers paid by the piece—proved to be a cost-effective way to acquire information from far-flung laboratories and from research specialties not represented among the expertise on the staff.

Slosson's outreach extended to both the famous (such as George Washington Carver) and the scientist in training.[34] Not every such interaction worked out well. Slosson complained that he had "spent the first month writing around to all my friends begging them on bended knee to write popular science articles for us" and then "spent the second month sending the articles back and telling them how rotten they were in such polite language as to induce them to send soon some better ones."[35]

WATSON DAVIS ON THE DOORSTEP

Years later, Watson Davis would say that he had been waiting on the front steps when Slosson arrived for his first day at work. Davis probably meant the anecdote as an expression of his enthusiasm for Science Service, and he did turn out to be a superb fit. Son of a Washington, D.C., high school principal, Davis had earned a civil engineering degree at George Washington University and was working on the research staff at the National Bureau of Standards testing concrete while moonlighting as a science writer for a local newspaper. Although his literary skills paled in comparison to Slosson's, Davis had both the instincts of a journalist and an engineer's ability to organize tasks. He could chase down news, glean the essence from dull research reports, and edit with feral skill; he also proved to be an efficient manager.

FIGURE 5. Watson Davis, managing editor of Science Service, in Quebec, Canada, with the Princeton University Geological Expedition, August 1927. Davis regularly joined scientific excursions, playing a dual role as participant and journalist. Courtesy of Smithsonian Institution Archives.

Sometime before January 22, 1921, Slosson met the twenty-four-year-old Davis and suggested that the aspiring journalist submit a list of the subjects that he was "more or less fitted to handle," along with story suggestions. Davis sent a long list that included "Subjects in which I have more or less special knowledge and training," such as civil engineering, building materials, geology, mineralogy, paleontology, ceramics, architecture, and ("with the assistance of Mrs. Davis") chemistry. Helen Miles Davis, Watson's wife and college sweetheart, had a B.S. in chemistry, and she later became involved in the organization's work as a writer and editor of the magazine *Chemistry*. Among other subjects he could handle, Watson listed virtually every other scientific discipline. The letter exuded the self-confidence of a good journalist. He did not purport to be expert in these topics, but he knew he had the intelligence to locate information and the ability to translate scientists' descriptions and interpretations. He was right.

In February 1921 Slosson wrote to Davis that Science Service was starting a weekly syndicate sheet composed of short articles on all sorts of scientific topics. Davis was encouraged to send material "as frequently" as he could. "How many we can use and how much we can pay for them will depend on how many editors select your stuff from the sheet," Slosson noted, adding that "this will not be a bother to you because you can often use another phase of the same paper or address that you are writing up for the Herald."[36]

For at least the next year and a half, Davis remained on the full-time staff at the National Bureau of Standards, while Slosson worked closely with Howard Wheeler, who supervised the offices and finances. Although Davis took on ever more tasks for Science Service, he was not presented as a regular member of the staff until much later. When a trustee complained that most stories seemed to be emanating from Washington sources, Slosson explained that "we are using the part-time service of Watson Davis, a young man of the Bureau of Standards, who has been running a successful science column in the Washington Herald."[37] That is to say, the name and role of Watson Davis was not yet known in September 1921 to one of the organization's most prominent trustees. As late as February 1922, Davis was listed as Science Service "news editor" but was still working at the Bureau of Standards and filing stories for the *Herald*.

The situation changed significantly in the next few months. In January 1923 Wheeler was fired, the position of business manager was abolished, and Davis was appointed as managing editor. By January 1924 a Scripps senior executive was saying "That young man whom Dr. Slosson engaged as Editor—Watson Davis—is a 'star'. He seems to me to be exceptionally capable, as well as enthusiastic, vigorous and very likeable."[38]

Davis, in fact, had an essential attribute for success in the news business, similar to qualifications that Scripps had once described:

> The first qualification for every member of my staff would be that he should have a "nose for news"! Every experienced reporter or general newspaper man

is quick to learn that the source (the man or woman) of his best articles is never (or at least seldom) aware of the relative importance or value in a news way of what he communicates.[39]

Cultivating such news sense was especially important for the science journalist because, Scripps noted, "the greatest handicap that any scientific man must bear in pursuing journalism is [that] he knows so much that he presumes that other people are far more largely informed than they really are."[40]

In addition to a well-qualified staff, the organization benefited from endorsement by the most prestigious scientific associations, from the advice of a talented and well-connected board of trustees, and from a favorable location. Early in the planning stages, Scripps and Ritter had engaged representatives of the American Association for the Advancement of Science, the National Research Council, and the National Academy of Sciences (NAS), a triumvirate that Ritter believed would provide a perfect balance of "bourgeois" and "proletarian" science. The NAS, he wrote, "represents the *standardized* scientific excellence" while the AAAS is "an open, free-for-all body" that has as its potential membership every scientist in the country; the NRC would serve as "a sort of suspension bridge" between the two.[41]

At the first official meeting of the trustees on May 20, 1921, five groups of people were present. Ritter, Scripps, and Scripps's son Robert P. Scripps served as the official representatives of the Scripps estate. Three other men were appointed as "representatives of the journalistic profession," including the president of the New York Evening Post Company and the well-known editor of the *Emporia Gazette*, William Allen White, who was a friend of Slosson's from his University of Kansas days. Scientists from the NAS, AAAS, and NRC included Nobel physicist Robert A. Millikan, astronomer George Ellery Hale of Mount Wilson Observatory, John C. Merriam of the Carnegie Institution of Washington, and James McKeen Cattell, the powerful editor of *Science* and *Scientific Monthly*. By tradition, one of the NRC or NAS appointees always included the head of the Smithsonian and the current president of AAAS. The involvement of such distinguished and accomplished men sent an important signal. If they endorsed Science Service's mission and approach, then the rest of science must pay attention.

The location of the organization's first offices—within the National Research Council building at 1701 Massachusetts Avenue, N.W.—also signaled status. In May 1922 the operation was moved temporarily to 1115 Connecticut Avenue, N.W., but when the NAS and NRC moved into their impressive new building on Constitution Avenue, N.W., in April 1924, Science Service moved into offices on an upper floor and remained there until 1941.

PRODUCTS AND ACTIVITIES

All of the staff, trustees, and founders were convinced they were engaging in "nothing less ambitious . . . than the education of the American people in science."[42] Slosson's personal vision of how to accomplish this goal incorporated the approaches of both journalism and science. Professional writers would do the actual work—we "cannot expect the leaders of research to spend their time on popularization," he explained at the first trustees meeting—but these translators would not be "outsiders." They would be trusted surrogates with special access and preferably with extensive scientific training. Slosson explained to Frank Richardson Kent, editor of the Baltimore *Sun*:

> Science Service . . . is in the embarrassing position of the middle-man, a translator from scientific jargon into journalese. Naturally, we get kicks from both sides, from the scientists because we distort and degrade their discoveries, and from the editors because we are too high-brow.[43]

Throughout its first few decades, Science Service exploited that middle ground to advantage in two types of activities. One activity was subtle, uncompensated, and in effect an "expense," in that it diverted resources from the production of salable news products—and that was to help the scientific community in promoting the cause of science. This activity was vital to the organization's acceptance and also cultivated the goodwill that later enabled it to tap prominent scientists for the radio show. Within a few months of its establishment, the organization began providing public relations assistance to the Washington scientific community. In spring 1921, the visit to Washington by two international celebrities (Albert Einstein and Prince Albert I of Monaco) coincided with the NAS spring meeting. With the assurance that NAS would reimburse expenses, Slosson and Davis handled all press arrangements, securing abstracts of the scientists' papers in advance and then retyping and mimeographing them for distribution to 225 newspapers. They brokered a special illustrated article by Smithsonian scientist C. G. Abbot to their newspaper clients, arranged a translation of Einstein's speech, and distributed both the speech and Smithsonian secretary Walcott's welcoming address to press associations and journalists on the day of delivery. In advance of the prince's visit (to receive a medal for his contributions to the science of oceanography), they obtained and distributed his speech and multiple sets of photographs of the colorful monarch. In subsequent years, Science Service played similar roles in publicizing the visits of scientific celebrities and even, during John T. Scopes's 1925 trial for teaching evolution, served as liaison between the scientific community and the teacher's defense team. But this assistance was always provided privately, behind the scenes; several times Science Service declined requests to take over the American Chemical Society News Service, because that would involve "publicity work."[44]

The other activity—the organization's major work—was to produce accurate news reports about science in all possible formats and forums useful to the press, accessible to the public, and salable at a fair price. Within its first year of operation, it began to sell two primary products. The syndicated weekly news service, initially called *Science News Bulletin*, began on April 2, 1921. Mimeographed sheets of news stories, mostly three hundred to five hundred words in length, were mailed to newspaper clients who could choose to reprint any or all of the material. By October 1921 the *Science News Bulletin* service had more than thirty paid subscribers with a combined daily circulation of 1.5 million. Slosson reported to Scripps that they were "getting quite a lot of scientific stuff into print one way or another, about three articles a day syndicated in the newspapers, besides special features for Sunday issues and periodicals."[45] The following September this weekly service became the *Daily Science News Bulletin*. On March 13, 1922, Science Service began publication of *Science News-Letter*, a weekly mimeographed magazine aimed at individuals, with Watson Davis doing much of the writing and editing. From 1923 to 1926, *News-Letter* circulation hovered around a thousand. Advertised as "A Living Periodical Text-Book on Science," the *News-Letter*, normally priced at five dollars per year, was offered at a reduced rate to AAAS members and schools.

For decades, these two products provided the backbone of Science Service's revenue. Their content also provided copy to reedit for use in radio news bulletins. During the 1920s, additional products included a matrix service (which provided papier-mache matrices of illustrations from which newspapers could cast metal stereotypes), a cartoon series focusing on economics and business ("Cartoonograph"), feature stories marketed to outlets like United Feature Syndicate, and syndicated columns like Charles Fitzhugh Talman's "Weather Fallacies," Frank Thone's "Nature Notes," and "News of the Stars" (an astronomy feature written initially by Isabel M. Lewis of the U.S. Naval Observatory). By covering hundreds of scientific conferences, the staff broadened their network of contacts and cultivated an important resource for radio interviews. After the December 1921 AAAS meeting in Canada, Slosson told trustee D. T. MacDougal (who was also AAAS general secretary), "Going to Toronto was a good move. We more than made our expenses and have been getting some complimentary letters from members whose papers we reported—and it is not easy to satisfy a scientist."[46]

NETWORKING, HOT TOPICS, AND MAKING NEWS

Economics and sensitivity to the marketplace—especially to newspapers' demand for timeliness—shaped content in print much as in radio. The Toronto AAAS meeting was the group's first attempt at "telegraphic news," wiring news stories directly to subscribers from the meeting. Slosson reported: "I wrote a thousand words a day which were telegraphed to nine papers and Mr. Davis contributed a shorter daily article to the United Press during the entire session."[47] Soon the mailed news reports were being regularly supplemented by telegraphed "special bulletins."

Nor did Science Service wait for news to cross the desk. Staff members began participating in eclipse expeditions or serving as local press representatives for such celebrities as Arctic explorer Knud Rasmussen and anthropologist Richard O. Marsh. One colleague suggested they might create their own news by having President Warren Harding "ring up King George by wireless telephone" and then publicizing the role of scientists in the mechanical arrangements for the call. Or they might secure an interview with Thomas A. Edison and ask him about "transmission of power by wireless to operate factories, locomotives...."[48]

Although Slosson approached neither president nor king, he and the staff did skillfully exploit many natural events. "The newspapers," Slosson explained, "demand 'news,' that is, something which has a definite event on which to hang the general information and necessary explanation."[49] Seismologists around the country were encouraged, through the Seismological Society of America, to wire information about major earthquakes. Eclipses provided other "timely" subjects for coverage: "An eclipse being one of the few scientific events of a spectacular nature which can be predicted in advance, we are going to take advantage of public interest to get as much material in the papers as we can about the sun and recent progress of astronomy in general."[50] In 1923 Slosson arranged with Walter S. Adams of Mount Wilson Observatory to join an astronomers' expedition to observe a solar eclipse, offering to assist in any way possible:

I don't know anything about astronomy of course, but I can be relied upon to do as I am told. I know that on eclipse expeditions in remote regions it is customary to draft in everybody, including camp followers like myself.[51]

Slosson's "exclusive coverage" of the eclipse was promoted to newspaper clients as "one of the most important events of the year":

Every one of your readers will be interested. Let a famous scientist and writer give them the news of this spectacular phenomenon. Dr. Edwin E. Slosson will report the eclipse from Point Loma, Cal., where dozens of astronomers will locate. He will be your special correspondent at very small cost.[52]

As a teaser Davis sent two "advance stories." While in California, Slosson also attended a scientific meeting in Los Angeles and wired back five-hundred-word dispatches for syndication. Income from that series more than covered his expenses. In addition to writing over fifteen hundred articles and another thousand incidental items every year, the staff had begun to develop a sophisticated "insider" approach to news gathering and news marketing.

TURNING POINT

The December 1923 AAAS meeting in Cincinnati symbolized an important turning point for Science Service, demonstrating that comprehensive and knowledgeable news coverage of a scientific event could generate positive publicity for all science. Harry L. Smithton attended the meeting as Scripps's personal representative and afterward wrote to Ritter:

I am sure you would have been greatly pleased to have seen the publicity given to that convention in ALL the Cincinnati papers. The leading headlines of the papers were given to the subjects of the convention and to interviews with the scientists. Collisions of the atoms displaced automobile and railroad collisions; slaying of bacteria and undesirable insects completely overshadowed similar "activities" among humankind; pictures of scientists ornamented the pages hitherto decorated by pictures of statesmen and criminals. Believe me: the scientist had his "day" in the way of publicity this time.[53]

And, he continued:

Not only were the Cincinnati Papers filled with science news, but even the newspapers in New York City gave big headlines and large space to the doings of the convention and interviews with scientists. Throughout the ranks of the scientists, Science Service received ALL the credit for this; and those who know what Science Service is have likewise given credit and praise to E. W. Scripps for giving to Science the greatest boon in a generation—connecting science up with the people, obtaining their interest and cooperation, thus not only assuring still greater advancement for science in the future, but assuring to mankind the great gifts that science would and can bestow.[54]

The news organization attracted important public recognition at the meeting, he noted: "The name of Science Service was brought before the Convention and it received direct praise and commendation as supplying to science one of the greatest needs of the day." Several of the trustees approached Smithton afterward and said, " 'Now here's something tangible—some tangible and tremendously important result of Mr. Scripps' work. Science has been placed before the people, and their interest aroused, as never before.' "[55]

Smithton told Ritter that the trustees' meetings "were a revelation"—he had been impressed by the participants' intelligence and their support for the idea of public dissemination of scientific ideas. Smithton's own enthusiasm for public communication of science was also significant. As a successful executive in the world of newspaper publishing and one of Scripps's most trusted financial assistants, Smithton recognized the essence of Science Service's accomplishment: convincing editors that this science news was especially trustworthy because it had been verified by scientists. The organization's not-for-profit status also forced responsiveness to the marketplace:

> Not only must its output measure up to a full standard of accuracy, but nothing goes out that is not first verified by men of science qualified to give such verification. . . . It can afford to actually produce not only well-written copy, but copy that has first been verified—that is authentic and UP-TO-DATE.[56]

Because the scientists themselves had become more engaged in supporting the work of popularization, Smithton concluded that "It is a movement WITHIN science, not a parasite endeavoring to fasten itself upon science."[57]

Within three years, then, Science Service had assumed a central role in brokering science to the masses, encouraging better news coverage of science throughout the press, and developing a spirit of cooperative popularization. The next spring, Davis and Slosson exploited similar tactics as they persuaded noted astronomers, biologists, chemists, and physicists to step into a new public arena and, eventually, consent to be interviewed on the air.

CHAPTER FOUR

※

Cooperative Ventures

Facts—Interest!—Thrill!! Newspapers using Science Service have discovered that science is not necessarily dry as dust.

ADVERTISEMENT IN *EDITOR &*
PUBLISHER, JANUARY 30, 1926

GADGETS and widgets, synthetic fabrics and imaginative innovations were tumbling rapidly from science and engineering laboratories in the 1920s. The rush of consumer products fascinated Science Service staff members much as it did other young people. Watson Davis and his colleagues owned automobiles, phonographs, and wireless sets; they went to the movies; they purchased cameras and sometimes developed their own prints. Not surprisingly, they also experimented with innovative communication technologies in their work.

In the organization's first six months, E. E. Slosson and Howard Wheeler, lured by the prospect of reaching "millions of people in a short time with a very vivid presentation of scientific information," had initiated a project to produce motion pictures about science.[1] Negotiations over the first project with a commercial motion picture company fell apart, but Science Service eventually produced one film with the U.S. National Bureau of Standards. *Make It Yourself* told the story of "an ordinary American boy" who constructs "an efficient, practical radio receiving set for about $6.00," using "an empty oatmeal box, a broomstick, and other easily obtainable material."[2] Exhibitions of the film during 1922 were sponsored by forty-three newspaper clients around the country, each screening coordinated with the publication of feature articles explaining the science behind the gadget.

Internal debate about the movie project had been contentious, and arguments over control of content foreshadowed problems that would arise later in connection with radio. There was an intractable clash of cultures. The trustees rejected one proposed contract because it afforded them insufficient "guarantee of supervision, of control, and of veto of footage."[3] Eventually, the movie project was abandoned altogether because the trustees were unwilling to relinquish control to any outside producer, yet reluctant to invest the organization's own resources in production. Slosson later called the "motion picture game . . . the most complicated and disappointing that I know anything about. The intrigues and pretensions and mendacity are amazing."[4]

Two years later, Slosson and Davis seized an opportunity to use radio as a medium for popularization. Compared to film, radio production required far less expenditure of time and money. This made it more promising for a modestly conceived series of talks. For the first two years, Science Service staff also worked in partnership with the National Research Council, an arrangement that added prestige and legitimacy to their efforts. These circumstances, and Science Service's growing reputation for accurate, "authenticated" science news, became critical factors in persuading dozens of prominent scientists to sit down in front of a microphone. Radio, Science Service promotional literature declared, is "fast becoming part of our national life." Now radio also became part of the public life of science.

GOING ON THE AIR

In 1924, not long after he first spoke on the air during an American Chemical Society meeting, E. E. Slosson finally bought a wireless set. "After long and proud resistance to the popular movement, I have at last joined the ranks of the radio fans," he wrote to one acquaintance, "and am spending hours listening at the receivers attached to my little crystal apparatus."[5] Broadcasting fascinated even such erudite, well-educated scientists. "I heard your talk last Monday night over the radio and enjoyed it immensely," Slosson wrote to one university professor:

> It was the next thing to having you present in person and basking in your genial smile. Every word came over the radio perfectly and in the characteristic and recognizable tones of your voice.[6]

Slosson loved getting similar feedback. He thanked one friend, saying: "Such notes are especially appreciated since when one is talking by himself alone in the broadcasting room, he has a feeling that nobody is listening to him or that they have switched him off after the first five minutes."[7]

That spring, Science Service was thriving. The fledgling organization was in a comfortable financial position, receiving $30,000 a year from E. W. Scripps and earning another $30,000 from product sales. The Scripps donation was being invested in an impressive array of securities. Operations expenses and salaries could be met by pooling the investment interest with the sales income.

In April 1924 Slosson and his staff moved into rooms within the new building housing the National Academy of Sciences (NAS) and the National Research Council (NRC). This proximity made it easier for the NRC to request assistance when a new Washington station, WCAP (owned by Chesapeake and Potomac Telephone Company, a subsidiary of AT&T), invited the council to arrange a series of scientific broadcasts. The NRC quickly established a Committee on Radio Talks (consisting of Slosson, chemist H. E. Howe, and NRC staff members W. E. Tisdale and Albert L. Barrows) and began to plan a series of ten-minute talks to be broadcast every week at 8 p.m., starting in June 1924. Tisdale supervised most arrangements during the first season, working closely with Watson Davis.

Science talks were already being regularly broadcast elsewhere in the country under the auspices of the Field Museum in Chicago, the American Museum of Natural History in New York, the Smithsonian Institution, and various universities. The NRC/Science Service talks followed much the same format. The first talk, on "Insect Sociology," was given by Vernon Kellogg, the NRC's permanent secretary, on June 6; the second, "The Atom," by Paul D. Foote, a U.S. Bureau of Standards physicist. On June 18, C. E. Munro of the U.S. Bureau of Mines described "Blasting a New Face on Nature" and in July George R. Mansfield of the U.S. Geological Survey asked and answered the question "How Old Is the Earth?"

Slosson, ever the diplomat, worried about the perception of competition with the Smithsonian series, so he offered to coordinate programs with Austin Clark: "It is a big chance and there are so many new things coming up in all branches of science all the time that there will be no excuse for duplication."[8] Clark too seemed intent on assuring other Washington scientists that all parties would be "working out our respective programs in collaboration."[9] Within little over a week, however, Slosson was admonishing his guest scientists to choose topics for their "touch of novelty" and "general interest," and no one appeared to be worried about overlap.[10]

As had Clark, Slosson dreamed that the radio broadcasts would reach "millions." His first published announcement of the talks in the *Daily Science*

News Bulletin (May 31, 1924) predicted that, although the allotted time might be "short," the audience would be "large": "A hundred thousand or a million may be listening in, and to interest such an audience in scientific achievements is worthy of [the] best efforts in the preparation of the talk." If the talks were successful, Slosson told one potential speaker, then they might also be broadcast on AT&T's commercial station, WEAF in New York City, where they would "reach a very large number of listeners."[11]

Such thoughts were not that far-fetched. AT&T, WCAP's owner, could use its own telephone wires to transmit the Washington program to WEAF, whereas stations owned by rival companies (such as WRC, owned by RCA) would be forced to lease expensive telegraph wires to link their affiliates. Once the issue of wire hookups was resolved after 1926, the development of economically viable radio networks moved quickly. Until then, however, the NRC/Science Service talks could be "broadcast" electronically in other towns (although there is no evidence that this happened often), while Austin Clark and the Smithsonian could "rebroadcast" only by sharing printed talks.

With each potential guest, Science Service negotiated topics and titles with an eye to audience appeal. Slosson asked H. E. Howe, editor of *Journal of Industrial and Engineering Chemistry*, to include "graphic details that will interest your auditors, who are not familiar with modern glassmaking and the new inventions" because "Dr. Tisdale and I, in looking over this, agree that it is rather too highbrow for a radio audience, that is, you have attempted to make it somewhat too encyclopedic and comprehensive."[12] Geologist George R. Mansfield was asked to address the question "How Old Is the Earth?" on the assumption that that would be "the most interesting one for the general public."[13] In addition to shortening the proposed manuscript, Tisdale and Slosson told Mansfield to "round off" his numbers: "for instance, put the Cambrian at 'half a billion years' instead of '550,000,000.'"[14]

THONE COMES TO WASHINGTON

One of the promising young graduate students that Slosson recruited as a stringer during the first months of 1921 was twenty-nine-year-old Frank Earnest Aloysius Thone, who was studying botany at the University of Chicago. Slosson's first letter to Thone elicited an enthusiastic reply, so Thone began to contribute news articles and later manuscripts for a popular book series edited by Slosson. Thone soon found writing more satisfying than university teaching, and he joined the Science Service staff as senior biology editor in 1924, just as the radio series began.

FIGURE 6. Watson Davis, Charlotte Davis, and biology editor Frank Thone. The Science Service staff was especially close-knit, engaging in picnics and swimming outings to the Potomac River during the summer. The children of Watson and Helen Miles Davis (Charlotte and her brother Miles) were frequent visitors to the offices. Courtesy of Smithsonian Institution Archives.

Thone probably never imagined he would become caught up in broadcasting. He certainly must have regretted moving to Washington in mid-July rather than staying in Yellowstone National Park, where he had served as a seasonal park naturalist for several summers. Thone never handled hot weather well. He explained to one friend that the initials "D.C." stood for

FIGURE 7. Frank Thone at WMAL microphone. Thone had just joined the Science Service staff as a writer in 1924 when he was drafted as a news reader for the *Science News of the Week* programs. Courtesy of Smithsonian Institution Archives.

challenging climate variations: "In the autumn and spring, D.C. means Delightful Climate; in winter it means Dank-Clammy, in summer nothing but Devil's Cauldron."[15] Apologizing one July for being late in sending articles to a client, Thone explained, "I haven't started yet. Genius burns but dimly with the temperature at 99, relative humidity damn' near ditto. Hope both the atmosphere and my steamed wits will clear soon."[16]

Such comments distinguish a man who was intelligent, literate, and witty. He was also thoroughly committed to the mission of serving science through popularization, even if achieving that goal required working in an urban cauldron every summer. A native of Iowa, a devout Catholic, and the son of a German immigrant, Thone had graduated from the University of Iowa and then continued his scientific training at the University of Chicago. After completing his doctorate, he had worked for William E. Ritter in California (1916–1918) and, like many young men at the time, served briefly in the military toward the end of World War I.

Thone then accepted a teaching job at the University of Florida, spending summers at Yellowstone. In 1923, in addition to occasional news articles, he had begun writing a popular book on botany and zoology, brokered to publishers by Slosson, and had completed a booklet on plant life in Yellowstone. When Slosson suggested that Thone join the Science Service staff, he was receptive. His teaching load was heavy (he was in charge of all work in botany and bacteriology), and life in Gainesville remained "dull even for a young semi-recluse like myself." Thone confessed that he "had been casting sheep's-eyes in the direction of science popularization as a regular occupation."[17]

At well over six feet tall, Thone towered over most people, including the rotund Davis. He remained single all his life, and was an enthusiastic tennis player until age, weight, and asthma slowed his game. His most appealing trait, and one that endeared him to friends, colleagues, and professional correspondents alike, was an ability to see irony and humor in unlikely topics. When an Iowa woman wrote to complain about Thone's use of the word "muddy" to describe the Mississippi River, for example, he responded with two pages of proof, including hydrology data on comparative particle rates in U.S. rivers. He meant "nothing derogatory," he explained:

> It's my river just as it is yours, and I love it and am proud of it. I recognize its muddiness in a purely factual way, just as I make no bones about having a big nose or red hair. It's something that calls for neither bragging nor apologies.[18]

Thone's surviving correspondence testifies to a wide network of loyal friends who reveled in such wit and admired his intellectual and moral integrity. A year after joining Science Service, Thone went to Dayton, Tennessee, to cover the trial of John T. Scopes, after which the biologist helped to establish and oversee a scholarship fund for Scopes to attend graduate school at the University of Chicago.

Thone's enthusiasm for botany and its popularization was unbounded. He characterized his approach to nature writing as "old-fashioned," in that he did not "hesitate to moralize when I feel like it, nor stop at 'anthropopsyching' my plants," but his prose style was clear rather than flowery, and he strived constantly for accuracy in describing current research.[19] He also endorsed Slosson's socially responsible approach to popularization. Slosson had explained that a "special feature" of the book series he edited was its focus on "the influence of science on human life." In the case of Thone's volume on botany this would entail emphasis of "the close relation between scientific plant culture and genetics, the production of wealth, the type of industry and common life."[20] This cultural focus connected science to the lives and occupations of the books' readers, interpreting "science" broadly to include economics and other social perspectives. And it was emblematic of the Science Service approach to choosing guests and topics for its radio programs.

DELIVERING THE NEWS

In the summer of 1924, as the temperature rose along the Potomac, Austin Clark had suspended the Smithsonian radio talks and headed north to New England. In the Science Service offices, however, the work of delivering the news continued. Slosson and Davis knew that scientists tended to leave town during the summer, so they proposed to Tisdale that the program carry on under the title *Science News Talks* and only occasionally feature talks by experts. This suggestion initiated Science Service's dual approach to broadcasting—its staff appropriated radio as another outlet for the science news produced for syndication, while also arranging for scientists to be "visible" to the public through interviews or discussions of their work. Slosson argued for flexibility ("Whenever a prominent scientist comes to town, we can draft him in for a talk or two") but agreed that a news-based format made sense:

> We have the best possible material for this since Science Service gets out daily from twelve to fifteen hundred words of news matter in regard to recent discoveries and forthcoming announcements of scientific events, put in nontechnical style. From this material we can pick out the items of most interest to the public and explain them briefly.[21]

This suggestion exemplified Slosson's practice of recycling text, both his own and others'.

Science News of the Week assured its listeners that "the progress of science is a continual excursion into the mysteries. . . . The impossible is continually being accomplished."[22] Tisdale, Davis, and Slosson each took turns reading the news or introducing guest speakers. On August 20, evoking themes he would return to again and again in future broadcasts, Davis opened by declaring that the most interesting science news of this week comes from the most distant parts of the universe. It has the effect of making us realize what a very small speck our earth really is.[23]

After reminding listeners of the distances between parts of the solar system and how long it took light from Alpha Centuri to reach the earth, he established a mood of astronomical mystery and adventure:

Do not attempt to restrain your imagination while I tell you about the latest announcements from the Harvard College Observatory. Investigations there under Dr. Harlow Shapley, director, have been centered on the Magellanic Clouds, great masses of stars and other heavenly bodies far beyond the Milky Way. . . . The Harvard astronomers have discovered that the bright stars of the Magellanic Clouds are larger and brighter than any of the giant stars heretofore known to astronomers.

Then, after an ably translated description of Shapley's work, Davis turned to another popular topic: "Is there evidence of intelligent life on Mars?" Astronomers "are divided into two camps on this question," Davis explained, citing a recent talk by Heber D. Curtis, director of the Allegheny Observatory, and summarizing arguments on all sides. Most astronomers do not favor the potential of any life on Mars "higher on the scale than the fungi," Davis emphasized, and yet he could not resist tickling the imagination one more time. "Do you believe that Martians can signal the earth?" he asked the audience. "Martians would have to wave a flag the size of Pennsylvania before we could notice it," he answered, then elaborated:

No lights or other signals have been observed. It is true that the Martians have been blamed, unjustly, for messing up radio communication. It is even suggested that all radio stations remain silent for a few days during the close approach of Mars in order to give the Martians a chance to signal us. Radio experts are skeptical of the possibility of any signal from Mars reaching us or of any of our signals reaching Mars.

Canals on Mars, Martian maps, experiments in perception and deception— fascinating distractions for Washingtonians on a hot August night.

The regular series resumed October 1924, still under the title *Science News of the Week* but once again centered on scientists' talks, occasionally with themes of business or political interest. These were not programs aimed at young audiences; this was science for adults. The series took full advantage of work being done within federal agencies. C. E. Munro described Bureau of Mines work on explosives and, a month later, explained spontaneous combustion; his colleague S. C. Lind talked about scientific work on helium; H. L. Gilchrist of the Chemical Warfare Service discussed "Chlorine Gas— Its Use in War and Peace"; and G. W. McCoy of the Public Health Service told about the "Pneumonic Plague." In December university professors explained "The Relation of Chemistry to Medicine" (Marie O'Dea of George Washington University) and "How Scientific Discoveries Are Made" (Ernest Merritt of Cornell University).

The positive public reception to the series prompted Maurice Holland, director of the National Research Council's Division of Engineering and Industrial Research, to suggest that NRC and the National Academy of Sciences launch "a national campaign to popularize science" through sponsorship of biweekly radio talks by scientists and engineers.[24] The following spring, Albert Barrows forwarded Holland's proposal to Watson Davis, asking whether Science Service would be interested in cooperating on the project.[25] At the instigation of Austin Clark, Barrows also drafted a separate memorandum on "Broadcasting Scientific Information by Radio," pointing to the success of the NRC/Science Service and Smithsonian series as an argument for a more ambitious extension of efforts nationwide.[26]

Print still had more prestige among scientists, however, and Clark was especially interested in publishing versions of the Smithsonian and NRC/ Science Service talks in a "series of booklets." Clark pushed this scheme to Barrows and others, but Davis and Slosson declined to participate on the grounds that they were already planning a number of popular science books, to sell at fifty cents each, which would conflict with Clark's proposal.[27] Although Davis privately questioned whether talks prepared for radio would prove to be suitable and readable copy for books, Science Service soon began publishing edited versions, under the general title of "Radio Talks," in the *Scientific Monthly*. This practice of supplementing broadcasts with print continued through the late 1930s.

By spring 1925 Slosson had accepted that radio communication might be a regular popularization venue, but he believed the "radio craze" had reached a new stage:

> Radio will continue to be an increasing factor in modern life, but some of the first fine frenzy has worn off. It will become, like gardening and automobiling, a thing of permanent interest and importance.[28]

The arrangers had begun to broaden the scope of the series, with the hope of broadening the audience. Slosson and Davis identified potential speakers from industry and universities as well as government, and from beyond the immediate Washington area. They also sampled from a wider range of scientific disciplines, never forgetting that topics like birds, insects, geology, and meteorology were reliable favorites among listeners.

The year 1925 turned out to be a busy one for the entire Science Service staff. Slosson conducted a private interview with Thomas Alva Edison. Davis took his first trip to Europe, attending scientific conferences, meeting Marie Curie in Paris, and, with the help of a stringer, obtaining a brief audience with Albert Einstein in Berlin.[29] Davis and Thone also became involved behind the scenes in the "trial of the century"—the prosecution of high school teacher John T. Scopes for violation of Tennessee's antievolution statute.[30]

Like the Smithsonian programs, the NRC/Science Service broadcasts avoided direct discussion of the evolution controversy during 1925. Instead, the series emphasized practical subjects such as navigation, fixed-nitrogen fertilizers, industrial alcohol, earthquake measurement, and the spinning of "artificial silk"; grand geological questions like how mountains are formed or how much the earth weighs; or topics in astronomy, archeology, and paleontology. Archibald Henderson asked "Is the Universe Finite?," Henry Norris Russell described the "evolution of the stars," and geologists routinely estimated the earth's age in millions of years. No talks were scheduled on evolutionary biology—and no one mentioned monkeys.

This silence is curious in light of Science Service's participation that year in the defense of Scopes. The choice to ignore the topic probably reflected scientists' continuing uneasiness with public controversy. Certainly, the airwaves were filled that spring and summer with the rhetoric of antievolutionists and with debates between defenders and critics of Scopes. Although the scripts for that summer's broadcasts have been lost, the Scopes case was surely mentioned in the news bulletins read by Slosson and other staff members while Davis and Thone were in Dayton, Tennessee, to cover the trial, even if the Science Service radio series was not used as a platform for evolutionists.

The trial did provide an important lesson about the power of the new medium of radio when the Chicago Tribune Company persuaded the Rhea County court to allow its radio station's microphone and announcer into the courtroom.[31] Because of the cost of leasing telephone lines to transmit audio back to its Chicago base (estimated to be over a thousand dollars a day), the station focused on the trial's more colorful segments, such as sessions when antagonists Clarence Darrow and William Jennings Bryan were expected to speak at length. WGN's three microphones stood "like fence posts" in the courtroom, adding an anachronistic touch to the antievolutionists' fight against modernism.[32] The station reassured its listeners that the broadcasts would not be offensive. The trial would differ "from most criminal trials in that neither the testimony nor the arguments will contain matter that may not with propriety be conveyed to the most delicate ears."[33] This science would be safe for your living room.

In fact, the broadcasts resembled coverage of a prizefight more than of a scientific debate; the proceedings were narrated in a "play-by-play" format, prefiguring the "actuality" broadcasts that became standard fare in the 1930s. For a half hour before court convened, announcer Quin A. Ryan would describe "every detail of the assembling scene, the entrance of each character, the manner, the apparel, the view through the windows, the atmosphere." Ryan heightened the drama by "announc[ing] the name of each witness or attorney as he arises to speak" and, during recesses, interviewing visitors and Dayton residents to add "color" commentary.[34]

Just as modern "live" broadcasts on television can captivate viewers with a sense of immediacy, the broadcasts from Dayton upstaged the usual musical concerts and talks. As one radio critic noted, "The air was full of worthwhile entertainment last night, but there is only one thing that sticks, persists, almost haunts at this late hour, and that is W-G-N's broadcast of the Tennessee evolution trial." He called the programs "deeply impressive," "with considerable human interest incidents and several comedy touches, as all great dramas should have."[35] The Scopes trial not only put science on the front page; it demonstrated that science could be an effective backdrop for radio drama.

CHANGEOVER

After another summer break, regular Friday night broadcasts of *Science News of the Week* again resumed in fall 1925. The following spring Tisdale moved to Paris, and a new Science Service reporter, twenty-five-year-old James Stokley, took over direct supervision of the programs, working with Barrows.

"We believe that Science Service has a real and substantial future," Davis had written to Stokley upon offering the job. "It needs a lot of good, hard work and lots of new ideas." Davis then described the ideal staff member—in words that applied equally to himself and to Thone:

> [T]he ideal Science Service staff member is a person who can get the news, write it, edit it, sell it, and, if need be, run the mimeograph that multiplies it. . . . While

nothing concerned with newspapers has the security of position that a teaching job has, we feel that the future is stable and safe—it can be made or broken by our output and conduct.[36]

Davis was convinced that Stokley would fit this template, and his judgment proved correct. Stokley was a skilled amateur astronomer and a graduate of the University of Pennsylvania. Although he left Science Service in 1931 to become director of the Fels Planetarium, he continued to write its regular syndicated astronomy column, returned to the staff briefly before World War II, and kept up a cordial relationship with the organization thereafter. Davis described Stokley as "an ambidextrous chap, having a graduate degree in psychology, teaching biology at Central High School in Philadelphia, taking advanced work in physics, being an amateur astronomer, having served as a press photographer in Philadelphia for several years, and writing popular science on the side."[37] This resourceful young man joined a full-time staff of fifteen, which also included several dynamic women journalists.[38]

After taking over from Tisdale, Stokley informed WCAP's station manager that the series would continue during the summer months, again centered on news rather than formal talks. Although Stokley planned to read most of the scripts himself, he emphasized to the station that the program remained a corporate effort and should be credited to Science Service, rather than to any single individual.

In June Science Service began to syndicate its scripts nationally, offering to mail the text of *Science News of the Week* without charge to stations around the country. Each script would be "written so that it can be of interest and comprehensible to the majority of listeners of the average station" and could be read on air in fifteen minutes by a local schoolteacher or announcer. Pronunciation of uncommon words would be indicated on the script. By midsummer about twenty stations, most owned by universities or newspapers, were receiving the mimeographs about a week before the scheduled broadcast date.[39] In another opportunistic marketing decision, Science Service also began to offer free mimeographed copies of the talks to the audience. Whenever a listener wrote for a copy, Science Service gained the name of a potential subscriber for *Science News-Letter* and would send a sample issue or other promotional material along with the script.

Topics were chosen to coordinate with the illustrated feature stories going out to newspapers. The first such printed script, intended for broadcast on July 6, 1926, discussed poison ivy and tied in with the article "Poison Ivy Yields to Science," mailed to clients for release July 4. The third printed script, "Unearthing an African Empire—Carthage," described an archeological dig and was coordinated with the print feature "Ancient Carthage Is Scene of Real Estate Boom." The radio program *Science News of the Week* thus became a "weekly summary of scientific news" based on material sold as part of Science Service's other products—the *Daily Science News Bulletin*, feature articles, and regular columns like "Star Story Map" or "Why the Weather."

This pattern of interlocking use of material emphasizes the repetitive, reinforcing nature of popular science content. Material published or broadcast in one forum was usually based on material published or broadcast elsewhere. Moreover, popularity drove content. If stories about research on poison ivy sold, then the Science Service staff wrote more like them, other news outlets ran stories about the research, and discussions appeared in multiple contexts. Unlike the lemminglike coverage of "hot news" today, where attention clusters briefly, peaks, and fades, the coverage of science topics in the 1920s and 1930s expanded more like a spider web, looping back as journalists discussed new applications of old knowledge.

UNCERTAINTY

That summer the broadcasting industry in the United States underwent its own complex expansion, entering a phase of dramatic reorganization that would eventually reduce the amount of airtime available to educational and similar noncommercial programming. The first hint of change occurred in late July when several local stations changed hands. WCAP, the station carrying the Science Service programs, was abruptly sold to Radio Corporation of America (RCA), which already owned WRC, the station carrying the Smithsonian and National Zoo programs. On July 28 Stokley met with WRC's program director and agreed to present the weekly *Science News of the Week* programs through that station.

At first the summer series continued as before. On August 6 George R. Mansfield of the U.S. Geological Survey told "The Story of the Northern Rockies," and the next week Frank Thone discussed a timely topic, "Hay Fever." The regular guest speaker series was scheduled to resume on October 5, 1926, and Stokley began planning programs around the first public demonstration of W. D. Coolidge's cathode ray tube, which was to take place in Philadelphia on October 20. There was a feeling of optimism about the radio work, tempered with awareness of its ephemeral nature. Slosson boasted that the broadcasts and script service reached

> a potential audience of millions, though it is always a question how many of these potentialities are "listening in". When I talk into the microphone I have

the discouraging impression that everybody has switched off and I am talking to myself in the studio.[40]

On October 18 the National Research Council's executive board decided to withdraw as coproducer of the series on the grounds that the "experimental" phase of "this new mechanism for the popular dissemination of scientific knowledge" was ended.[41] Science Service could manage any further development on its own.

This was an amicable change, but NRC's withdrawal reflected growing anxiety about who really controlled the broadcast content. The old relationship with WCAP had been cordial, but WRC was tied more closely to its parent corporation RCA, which was then in the early stages of developing a national network. The NRC Committee on Radio Talks noted that WRC "finds itself less at liberty to receive local contributions on account of linkage with a supply net centering in New York City."[42] Moreover, WRC had an existing relationship with the Smithsonian and National Zoo. Science was being squeezed in the schedule.

Several months of uncertainty followed, during which Science Service remained firmly committed to radio. In November 1926 over a dozen stations were using the syndicated *Science News of the Week* scripts. As changes in the broadcasting industry rippled around the country, stations that had taken over other local stations wrote to add the talks. Then WRC decided to reduce the sustaining (noncommercial) time made available for science. The station continued the Smithsonian and National Zoo talks and left Science Service in the cold. Stokley quickly arranged to move *Science News of the Week* to another local station, WMAL.

When Clark learned of the reorganization in the local market, he initially greeted the situation with optimism. "His" station would be expanding its schedule, he wrote to various acquaintances, and this change could mean more exchange of material. For many months he continued to insist that the pressure on noncommercial time would not last.[43] Clark had already arranged that, during the coming season, Smithsonian programs would be relayed by wire once a month to Boston and New York City stations, and the Thornton Burgess talks would be broadcast in Washington. Then in August the Washington station began to balk at the $75 connection fee. By September it was becoming clear that WRC would not continue all the science talks that it and WCAP had carried the previous season. Because WRC would be favoring commercial clients, the amount of time available for sustaining programs could actually be less.

Even though Clark compromised and agreed to combine his original series with the zoo talks, the situation remained uncertain. In early November

Clark confided that the situation was "chaotic"—"Temporarily . . . science is in abeyance."[44] He waited to arrange the series until November, when William Mann returned from Africa and they could jointly discuss the matter with the station. On November 15 Clark wrote to Burgess that the "radio situation" appeared to be "clarifying":

> Station WRC called up this morning, and we go on the air Wednesdays at 6:45. . . . Science Service has been turned over to Station WMAL. I thought that if I just left them alone something of the sort would happen. The choice of Wednesday, which was made by the station, will make a hook-up with you much easier to arrange in the future.[45]

On the first Smithsonian program that fall, Mann described the colorful process of "Bringing Home Live Animals from Africa." He had returned with enough specimens to more than double the population of the National Zoo, including two young giraffes who were nine feet tall and, as Clark wrote, "extremely affectionate": "Affection is all right in the abstract; but when it is expressed by liberal application to one's face of a long wet tongue it has its disadvantages."[46]

The fifth (and final) season was under way. Between November 24, 1926, and June 29, 1927, Clark and Mann arranged twenty-nine broadcasts. The next fall, however, Clark was informed that the station would no longer carry any Smithsonian talks. He was off the air.

ADJUSTMENTS

Meanwhile, on WMAL, Davis and Stokley altered the Science Service format, interspersing news programs with occasional "talks by prominent scientists" and shifting away from the more conservative tone that the NRC had favored. The broadcasts were described as "a continuation of the series which Science Service and the National Research Council formerly presented through station WCAP."[47] Dignity remained important, but Stokley and the others began to experiment with topics and titles designed to attract more listeners. Frank Thone, for example, spoke on Thanksgiving Day on the topic "New Turkeys for Old."

The organization had a powerful incentive to continue, even under constrained circumstances. Radio may not have earned revenue for Science Service, but the broadcasts offered valuable publicity and the scripts gave opportunities for local newspaper tie-ins. The mailing list for scripts fluctuated between fourteen to twenty stations. Because these stations were sometimes owned by the same newspapers that purchased the syndicated products, the free scripts were an inexpensive way to add value and build client loyalty.

As scientific associations, recognizing the potential public relations value, invited radio coverage of their meetings, limitations on the time available for non-entertainment programs led to competition among groups. When AAAS met at Philadelphia in December 1926, both Stokley and Austin Clark (the association's volunteer "news manager") attempted to arrange broadcasts through local stations. Stokley proposed to Clark that Slosson speak on December 31, and suggested other speakers and various wordings regarding sponsorship. Clark took umbrage and claimed interference with the AAAS arrangements: "If Science Service insists on taking over for itself three periods offered the Association by Station WOO it will be necessary to cancel the invitations sent out by the Association."[48] Stokley replied soothingly and agreed that the talks should be announced as being presented "under the auspices" of both organizations.[49] Similar jockeying between Clark and Stokley marked the preparations for radio coverage of subsequent AAAS meetings.

When stations began to attract more advertising (so-called toll broadcasting), they sometimes stopped using the Science Service script service. Nevertheless, the scripts continued to be a popular product. By November 1927 more than twenty stations nationwide were subscribing, including stations in Texas and Oregon and WABC in New York City; a year later, the list had grown to thirty-one. When a decision by the Federal Radio Commission had the effect of decreasing WMAL's range and power, and hence the potential audience for the local Science Service broadcasts, the script service helped to keep the organization's radio outreach national.

AUTHENTICITY, APPEAL, AND SALABILITY

In choosing material for radio productions or for print, Science Service considered three goals: authenticity, audience appeal, and salability. "Timeliness," the usual standard for news stories, did not apply to science news as it did to reports of train accidents or house fires. No one could pretend that research like Shapley's work on Magellanic Clouds happened "overnight." Timeliness came to be judged instead by whether a result had been previously reported in the popular press, not whether the conclusion was news to scientists. The goal was to explain how each piece of information conformed with (or overturned) conventional wisdom within science.

"Off the beaten tracks real news is breaking," a Science Service advertising circular read. "What scientists and engineers are doing today will affect the world tomorrow. Are you getting this news?"[50] Radio's immediacy worked to advantage here. Science news presented dramatically—as Davis had done in the "life on Mars" segment—created an illusion of freshness even if the ideas were actually well-worn and well-known to specialists. Moreover, as Smithton and other newspaper professionals had recognized, a source of trustworthy, verified copy—reliable information that journalists might otherwise not be able to access or authenticate—could save clients time, money, and potential embarrassment. The certification offered by a Science Service byline, then, became one key to successful marketing. The organization's print articles and radio scripts could be trusted as accurate descriptions of research, no matter how abbreviated. Each scientist interviewed would be a respected member of his or her field.

On the question of what would appeal to an audience, by contrast, the perceptions of scientists and media executives (both broadcasters and newspaper editors) differed sharply. Scientists' notions of what would interest the public were shaped—then, as now—by what they themselves found interesting and appropriate. Because Science Service was seeking to market its products, though, its definition of "salability" acknowledged the interests of its newspaper clients and the recipients of its scripts; that is, its assessment of appeal, even for material provided free of charge, drew on pragmatic evidence of which stories sold and which did not. Reliable and authentic science that was interesting to other specialists might not necessarily be of interest to the general public. Science Service could not afford to pay a stringer for material that could not be brokered successfully to a client.[51] It was not sufficient simply to desire to disseminate science; success required becoming skilled in the art of selecting topics with popular appeal.

How then to fascinate the audience without pandering? The scientific community's disgust at newspaper sensationalism was undisguised. Some scientists took offense at any mention of "timeliness" (or, as Slosson phrased it, the "editorial superstition . . . that nothing is useful tomorrow that might have been published today").[52] Slosson himself regarded the notion of "spot news" as an "absurdity in scientific matters." Science exemplifies a "slow and orderly progress," he wrote. "The telegraph is rarely needed except in such cases as the appearance of an unknown comet."[53] Although he gradually came to endorse the notion of "science-by-wire" (news sent to newspapers via telegram) and to encourage development of networks of scientists who might telegraph information about seismological events, he recognized that Science Service would have to present the news of science sometimes *as if* that news were "new," when in truth most "scientific news that appears in our dailies is often two or three years old and sometimes two or three centuries old."[54]

Confident in the reliability of their products and the integrity of the production process, the staff enthusiastically adopted the techniques of two relatively new professions—advertising and public relations. Davis, an eager modernist, cultivated connections with industry public relations specialists, built networks of university publicists who facilitated access to professors, and established good relationships with advertising pioneers like Ivy Lee. As an engineer, he was comfortable, perhaps even enamored, with the industrialists of science at DuPont, General Electric, and similar companies, and he seems also to have understood how advertising and public relations could be used to advance the mission of civic education. In the late 1920s, Edward L. Bernays argued that "manipulating public opinion" by providing good information to consumers could actually help safeguard the public against its own "aggressiveness."[55] As Davis became friends with Bernays, he absorbed such ideas and incorporated them into his organization's work. Science Service was, after all, engaged in its own public relations campaign to persuade both press and the scientific community to join in educating the masses about science.

In the view of scientists and their organizations, only scientists had sufficiently high intellectual standards to assess accuracy, a view Clark shared. You "cannot develop a high standard from mediocre beginnings," he had emphasized. "Constant and rigid supervision of the output" is therefore necessary "to insure and maintain a high standard."[56] Balancing this commitment to accuracy with the need to sell products in the marketplace represented an important source of tension. At one meeting of the Science Service trustees, in September 1925, an argument erupted over what some perceived as a compromise in these standards. C. G. Abbot, assistant secretary of the Smithsonian, later apologized for his outburst:

> I do feel . . . that since Science Service is fathered by the two great scientific bodies we should try as hard as possible to put accuracy over salability, and yet I know that salability must, after all, come first or the experiment will fail. It is a Scylla and Charybdis proposition.[57]

Not all prominent scientists agreed with Abbot. Some regarded attention to the "salability" of science news as inappropriate. AAAS permanent secretary Burton E. Livingston, for example, could not hide his disdain when he wrote to Davis:

> One aspect of Science Service is commercial, for it aims to make its living by selling things, and you must necessarily approach a potential subscriber in a salesmanship manner. Our emanations [at AAAS] to members are on another basis: I try to keep salesmanship out of it altogether.[58]

Time and again, the editors at Science Service complained of the difficulty of balancing these goals, and they often expressed empathy with the scientists' complaints. When Frank Thone wrote a news story headlined "'Modern' Insects, Fast Livers, Like Bright Lights and a Hot Time," based on an Ohio State scientist's article in *Ecology*, he asked the professor for understanding:

> I hope you will not be too much pained at the way I yell for the editor's attention in my headline: This business of trying to make science compete for news space in a field already crowded with crime, "S.A." [sex appeal], and jazz generally presents its own problems![59]

RADIO AS MIDDLE GROUND

By the late 1920s much of the early criticism of Science Service had been defused. The organization was proving to be a credible interpreter of science on radio as well as in print. Nevertheless, distrust of all the press persisted within the scientific community, and with some justification. As journalist Philip Sinnott observed in refreshingly honest language, scientists and other academic experts sometimes did have cause for complaint:

> Imagine patient, dull work for years, in the interest of humanity, announcement of a discovery—and read what crackpot writers glibly tell of its power to make dwarfs gigantic, cretins highbrow, the fat lean, the lean fat, the poor wealthy, etc., including effect on Black Bottom dancers. Our academic professor is shocked, nauseated, humiliated. When letters begin to pour in giving private symptoms and enclosing money for his "medicine," he's damned angry at a press he believes has made him a charlatan.[60]

Others peered more deeply into what happened when science received such publicity and recognized that reporters were not the only ones at fault. When William E. Ritter read a copy of Sinnott's remarks, he commented that researchers are "not really as indifferent to the general significance of their work as they suppose they are" and most have probably not "analysed searchingly their own motives in research on the general nature of natural knowledge and its relation to mankind."[61]

In its radio programs as well as its print products, Science Service interceded between these two demanding constituencies in an effort to serve a third: the public. The "spokesmen and popularizers" stood between the scientists "absorbed in their specialties" and "the great mass of readers."[62] In January 1926, at the Medill School of Journalism, Slosson made an eloquent plea for the importance of the science journalist as "middleman." Popular indifference to science, he said, should not be ascribed to the educational system but to that system's failure to produce "competent and zealous interpreters of scientific thought.... The number of persons who can profitably devote their lives to research is relatively small in each generation, even though it ought to be larger than it is now in this country."[63] Scientists do not like people looking over their shoulders while they work, nor do most want to take time away from their research to engage in popularization, and so they would benefit from more "middlemen of science."

Although Science Service devoted most of its enthusiasm and efforts to perfecting this middleman role through print publications, it was actually its radio series that brought science to "the great mass" of Americans. Its ability to accomplish that task was due in no small part to the organization's financial independence. The issue of "who pays for popularization" proved a crucial factor in the amount and type of programming that began to appear on the air. Money shaped the scientists' attitudes toward public outreach, just as it drove the development and regulation of the U.S. broadcasting industry.

———— ✳ ————

CHAPTER FIVE

Shifting Ground

How long will it be before as large a percentage of the population crowd around radio loud speakers to hear a lecture—any sort of a learned lecture—as now "listen in" on the prize fights? Oh yes, I was one of the latter! I must be a low-brow, too.

HARRY L. SMITHTON, 1927[1]

WITH radio, "the whole country will join in every national procession," NBC executive David Sarnoff predicted in the August 14, 1926, issue of the *Saturday Evening Post*. Ordinary workers will attend "night school" via radio, and scientists will demonstrate their latest discoveries. Similar claims that radio would be an "instrument of direct education" rippled through other political rhetoric and social commentary.[2] Like the inflated promises accompanying many new technologies through the centuries, these proved to be dreams realized only in part—and never for all.

The failure of American radio to achieve such admirable goals, especially for civic education in science, can be traced to several causes, some related to the commercial nature of broadcasting and some to prevailing attitudes among scientists and science educators. Money drove the system. During the 1920s, radio evolved from experimental technology to potentially profitable business, from a playing field open to amateurs to a closed shop controlled by national commercial networks. During the same month that Sarnoff so enthusiastically touted radio's potential, he and executives at AT&T, General Electric, RCA, Westinghouse, and other major broadcast companies were secretly negotiating to carve up the industry. In exchange for the "exclusive right" to provide network interconnection over its telephone lines, AT&T

sold its stations and relinquished broadcast transmission and content production to its competitors.[3] In Washington, D.C., the former AT&T station WCAP closed down, WRC began broadcasting every day of the week, and the science programs carried on these stations were shifted to less favorable times or canceled. As the Smithsonian's annual report noted, with characteristic understatement, "the local radio situation became somewhat involved."[4]

The effects of the broadcasting shakeup were felt nationwide. Throughout the country, NBC and CBS were assembling networks of affiliates, and local station managers were losing control of programming. The national broadcasters themselves began to define the "public interest" and to decide whether content like science would serve that interest. These business changes and the political decisions that sanctioned them created a mass communication system dominated by what Thomas Streeter has characterized as "selling the air."[5]

For popularizers of science, the situation posed a quandary. Their concerns centered not on making profits but on maintaining control over the accuracy, quality, and timing of content, and ensuring that proper intellectual credit was given. With only a few exceptions, scientists did not subsidize (or, as a group, invest in) popularization of science. The success of Science Service had offered a temporary solution: a generous donor with deep pockets and the establishment of a professional organization to handle the promotion of science news. And while Science Service, Inc., thrived, the scientific societies saw no compelling need for additional investment on their part. Money, in fact, initially flowed the other way. James McKeen Cattell, editor of *Science* and *Scientific Monthly*, for example, accepted a generous stipend from E. W. Scripps for serving as a trustee. Science Service also frequently handled the scientific associations' publicity and press relations for free during the 1920s. The popularization of science was being reshaped, however, by more than scientists' willingness to cooperate or volunteer their advice and the number of exciting discoveries celebrated on the front page. Popularization was increasingly affected by the health, structure, power, and values of the commercial media businesses that delivered such content and by the national economic and political forces that swirled around them. In the late 1920s and 1930s, all these contexts were undergoing profound shifts.

The transformation of American broadcasting—which occurred just as the scientific establishment was becoming comfortable with addressing the public via radio talks—refocused attention on the economics of popularization. Who should pay for interpreting science for the public and for advancing civic education in science—and why? To what extent should scientists cooperate with the advertisers underwriting commercial broadcasting? Each organization and entrepreneur answered these questions differently. Some rejected "commercialism" altogether; some embraced it, cautiously. Others favored civic education, whatever the venue or compromise. This disparity (and lack of coordination) in responses exemplified dilemmas that would haunt science popularization efforts for decades to come.

EDUCATIONAL BANDWIDTH

To understand the broadcasting system that emerged in the United States during the late 1920s and the regulatory decisions that shaped that system, one must look first at the unbridled proliferation of entities able to transmit signals to an ever-expanding pool of listeners. By the end of 1926, Mark Goodman points out, there were "15,111 amateur stations, 1,092 ship stations, 553 land stations, and 536 broadcast stations" in the United States alone.[6] During the next year, another three thousand amateur stations went on the air. Initially, only limited frequencies were available for long-distance transmission, and amateurs were not yet confined to their own bandwidths. The federal government allowed only 89 wavelengths to be used by the 536 broadcast stations, most of which were concentrated in large urban areas. Stations like WRC and WCAP in Washington actually alternated broadcasting days until 1926. Signal interference was common and reception uncertain. When E. E. Slosson appeared on WGY in Schenectady, New York, in early 1925, he told a General Electric public relations representative that complimentary letters had arrived from Illinois and North Carolina, but when "Mrs. Slosson tried to listen-in to my talk through WGY [she] unfortunately could not get the station clearly on account of static or interference of some sort"—despite the seventy-five-foot antenna atop their northwest Washington, D.C., home.[7]

Everyone agreed that something needed to be done to control the chaos on the airwaves. But, as many historians of broadcasting have documented, improvement was stalled by industry objections to reducing the number of possible stations or to limiting their range.[8] Such action, it was argued, would inhibit the marketplace just as radio was beginning to seem profitable. In the United States, the First Amendment guarantees everyone the right to speak, yet allowing everyone to crowd up to the microphone at once was impractical. How could access to the airwaves be rationed or regulated without impinging on free speech?

The resulting federal legislation, Goodman observes, attempted to balance competing interests by granting the public both unlimited access to "entertainment" and "the right to listen."[9] That is to say, government would not inhibit the purchase or ownership of receivers (which did happen in totalitarian countries) and therefore would not restrain Americans' access to content, but most members of the public would lose the right to speak over the airwaves at any time or frequency of their choosing, that is, to broadcast without a license.[10] Amateurs were squeezed into certain bands and their transmitter use was severely constrained. Any American could still buy a printing press and print her own newspaper or book, but not everyone would be allowed to use the airwaves to give a chemistry lecture or violin recital. Access to the platform, for all performers, regardless of merit, would be controlled by the entities granted the appropriate licenses.

Concurrent concerns about programming resulted in establishment of the Federal Radio Commission (which in 1934 became the Federal Communications Commission [FCC]). This agency would not just regulate the technical aspects of the spectrum and decide *who* might broadcast but would also oversee (mildly) what they would be able to say. For communicating about science, this situation proved disadvantageous. The behind-the-scenes agreement had concentrated power and programming control among a few national broadcasting companies. As history shows, these companies saw profit in entertainment rather than in civic education.

SCIENCE AS "PUBLIC INTEREST" CONTENT

It need not have turned out this way. Communications historians like Erik Barnouw, Douglas Craig, Sydney Head, Michelle Hilmes, and others have long emphasized that the federal government had other options for controlling broadcast technology. Radio broadcasts could have been regarded as a form of speech like the texts of books, magazines, and newspapers and therefore subject to First Amendment protection with only minimal government oversight.[11] Or radio stations and networks could have been assumed to be private monopolies, to be managed by their developers, investors, and owners with little public involvement. The United States might also have followed the British model, whereby the national government controls the airwaves and transmission facilities, and taxes or fees (perhaps on sales or ownership of radio receivers) support program production.[12] This type of arrangement has been widely credited with nurturing the British Broadcasting Corporation's strong tradition of public service content, including its award-winning attention to science. Instead, the American system developed along a course that served the best interests of commercial broadcasters. Today the FCC awards (and, rarely, denies) broadcast licenses to individual station owners. Stations and networks are considered to be commercial operations, commodities that can be bought and sold. As consumers, we purchase our own receiving sets. The marketplace limits what we hear when we turn them on. Government does not tell broadcasters what programs to carry, and with the exception of military censorship during World War II and occasional morality-based restrictions on speech, government regulators have rarely proscribed radio content.[13] The marketplace and the broadcasting system's owners and advertisers have set the controls.

This situation has not been overwhelmingly disadvantageous for consumers. Until the advent of delivery systems like cable television, satellite radio, and podcasts, Americans could access the majority of television or radio content for free. Advertisers (and some philanthropic underwriters) subsidized almost all production and transmission costs, and these advertisers presumably recovered their investments through increased sales of goods and services. Even today, the American radio and television broadcasting system relies on industry self-regulation. The federal government places few content requirements on the communication organizations themselves, save for regulators' occasional exhortations that broadcasters should serve the public interest or responses to public controversy over some entertainer's purported challenge to prevailing moral or social values. The content of the electronic mass media stretches across hundreds of stations and outlets— including Internet-based communication—and exhibits both astonishing diversity and occasional bursts of breathtaking creativity.

The system emerged gradually after the business shakeup that began in 1926, with many of today's organizations and corporate relationships established during the subsequent period of expansion. The National Broadcasting Company (NBC) was founded in 1926, Columbia Broadcasting System (CBS) in 1927. The various corporate entities that manufactured radio equipment or transmitted entertainment, such as the Radio Corporation of America (RCA) and American Telephone and Telegraph Company (AT&T), developed organizational structures that changed little until later in the twentieth century. The companies built loyal audiences of listeners who tuned in every night expecting to be entertained and informed; they established business approaches that insured steady revenue streams and brought them extraordinary profits. Neither the 1927 Federal Radio Commission nor subsequent government regulatory action and legislation could shift the momentum away from the emergence of private, for-profit broadcasting systems. By 1928 CBS and NBC were engaging in "chain broadcasting," developing what would become national networks of affiliated stations.[14]

The social influence of the American broadcasting system came from both its penetration and its interconnectedness. By the end of the 1920s about 40 percent of all American families had radio sets and in some cities, almost three-quarters did; by 1932 Americans owned almost half the radio sets in the world.[15] As listeners began to perceive radio as a "national entity" rather than just one station broadcasting from a single location, radio helped to unite a country that was geographically enormous and culturally diverse.[16] The networks assembled by CBS, NBC, and the Mutual Broadcasting System (MBS) soon controlled what could be heard nationwide. Each station became less a local, independent business and more an "outlet" for "vertically integrated and nationally centralized systems of program production and distribution."[17] The larger the network, the greater the power concentrated in the hands of those who controlled it. By 1936, Douglas Craig points out, the affiliates of CBS, NBC, and MBS (that is, 37 percent of all U.S. stations) "commanded nearly 93% of the nation's total transmission power" and could determine almost all of what was broadcast on American radio.[18] Their operations were based on delivering entertainment, news, sports, and similar content to large, undifferentiated, national audiences.

This centralization of control over content affected scientists' access to the microphone and their ability to insist on "dignified" presentations. The lack of government concern about such content influenced both the types of science-related programs the networks themselves produced and decisions by national networks and local affiliates about whether to carry programs produced by nonprofit groups like Science Service. The locus of decision making moved away from station managers, people familiar with community resources and tastes—such as the ones who first invited Charles D. Walcott and Thornton Burgess—to network executives who sought corporate sustainability and network uniformity. Amateur efforts dwindled. To survive, productions were forced to achieve a level of sophistication similar to that of other network programming. Everything had to compete harder for attention.

Radio's power came from its national reach. Its audience appeal came from creativity fueled by advertisers' dollars. The first stations, noncommercial ventures, had put little effort into persuading listeners to purchase products or services of any kind, other than the radio equipment itself. WEAF in New York City was the first U.S. station to sell time to advertisers in 1922, but many stations at the time refused to air commercial announcements.[19] When Harlow Shapley suggested to Austin Clark that a Smithsonian broadcast using one of Shapley's astronomy scripts might also promote a forthcoming book based on the Harvard Observatory talks, Clark refused, explaining that "unfortunately Station WRC is licensed as a 'non-commercial' station and everything that can be construed as advertising is barred."[20]

Once networks and stations realized they could make money from "selling time," that is, from selling advertisers the opportunity to reach potential buyers in their own homes, these policies were abandoned and the airwaves became saturated with advertising.[21] In 1927 the average number of commercial hours broadcast on NBC and CBS network stations reached fourteen per week, and that figure more than doubled in the next two years.[22] On December 16, 1927, NBC revised the rate cards for stations in its Red, Blue, and Pacific Networks; time on the air had become more precious. Advertisers were willing to pay from $120 to $600 per hour, depending on the city.[23] Between 1927 and 1930, advertising sales by U.S. radio networks rose from $3.8 million to $26.8 million; by 1934 annual sales had reached $72 million.[24]

Thanks to the expanded influence of advertisers, Thomas Streeter observes, the 1926 shakeup stimulated the birth of "new cultural forms" like soap operas and game shows, as well as "new habits of everyday life" such as regularly tuning in to a favorite program.[25] Sponsors usually had no interest in subsidizing programs that attracted small audiences. Americans increasingly turned their dials to clever entertainment that had been either produced by sponsors or fine-tuned by networks to fit the sponsors' preferences.

The shift toward advertiser-driven programming did initially create some opportunities for educational content. Unpurchased time, usually late at night or at other times when fewer people were listening, still needed to be filled with sound. Dead air would drive listeners to rival stations. The networks used "sustaining" (non-revenue producing) programming, such as that produced by scientific groups or government agencies, to fill those slots. This circumstance proved to be a double-edged sword, however. Because the programs were provided to networks at no cost, and were scheduled for the off-hours, broadcasters came to perceive science as peripheral "public service" material rather than content pertinent to their business.[26]

Another impact came from the way that advertising was shaping radio's presentation styles. From the beginning of the century, American advertisers had adopted a tone for selling products that "approximated the one-to-one relationship between salesman and prospect"; on the radio, this meant attempting to build an emotional relationship between announcer and audience.[27] Radio's intimate, informal communication style was quite different from science's studied rhetorical precision. Roland Marchand has pointed out that some businesses perceived radio advertising not just as a platform for sales but as a tool for cultural redemption of the masses. These advertisers chose to support programming that would uplift listeners and favored moral messages over the rational analyses and technical discussions central to science programming. Thus, while the networks' power grew, radio's "extraordinary possibilities" for education and enlightenment, including for civic education in science, were being only marginally exploited.[28]

From the onset, radio had been distinguished by its diversity of content—from astrology to politics, from "how-to" talks and practical information to every form of entertainment—but that too now changed.[29] Inadequate financial support for educational stations and the untrammeled success (and political clout) of the commercial networks led to the dwindling of educational broadcasting on a national scale and, therefore, to fewer outlets for educational material such as science.[30] Almost one-third of educational stations licensed between 1921 and 1937 "failed within their first year of operation"; during the Great Depression, many nonprofit station owners such as schools and churches sold out to commercial interests.[31] The federal government did little to reverse the trend. Indeed, Erik Barnouw and many other historians argue, federal policies actually accelerated the sell-off. When the Federal Radio Commission eventually introduced the phrase "in the public interest, convenience, and necessity" as part of the licensing requirement for broadcasters, the voices promoting education were weak. The clause was interpreted to mean that "radio schedules must include offerings of social value," such as news, lectures, classical music, or "programs designed to illuminate matters of special concern."[32] There was, however, no special emphasis on educational value or on science.

BURGESS CONSIDERS A SPONSOR

Radio's patterns were being set. Once WRC began to accept "paid-up advertising" in 1926, Austin Clark observed, there seemed less time available for science, even though the various Washington groups offered to combine their programs. The situation mystified Clark: "It does not seem to me that radio advertising is worth the terrific prices that it costs to commercial firms, and I do not believe that the present conditions can last very long."[33]

Nevertheless, sustaining programs that were considered to be "educational" (such as the Smithsonian talks) became increasingly vulnerable to cancellation. "My series here was not continued this year; the station did not approach me, and I did not make any advances to the station," Clark wrote Thornton Burgess during spring 1928. He characterized the decision as purely economic. The Smithsonian series, he explained, had attracted only "non-spending elements, especially old women and children," whom broadcasters regard as "negligible"—"The youthful spendthrift jazz consumer and the money-making man of extravagant habits represent those that they wish to please."[34]

Burgess fared somewhat better, in part because of his down-to-earth approach to broadcasting. He understood radio's commercial foundation and the importance of strategic compromise. When Clark urged putting on more programs, Burgess argued that "two scientific talks a week" per station should be sufficient to "keep the appetite always whetted."[35] That spring, however, Burgess also found himself in jeopardy when WBZ indicated it might drop the sustaining series. "I suppose," Clark offered in consolation, "the companies running the stations are all reducing their expenses to a minimum, gathering up all the paid programs possible, and in every way trying to cash in heavily on the present popularity of the radio."[36]

At first, Burgess did not seek commercial sponsorship. He had accepted voluntary donations in the past, and a partial subsidy could keep the series alive. In 1925 Connecticut Valley Lumber Company had donated a small sum "to be used in any way I see fit to promote the work of the Radio Nature League . . . because the officers of the company believe that the Radio Nature League is actually accomplishing something for conservation."[37] If WBZ lost interest, Burgess was confident that he could switch to a nearby station owned by the *Hartford Times*. Then "a good angel" came forward, allowing sufficient subsidy for Burgess to continue on WBZ.[38]

As the Radio Nature League began its fifth year, Burgess thought about canceling but listeners always urged him to remain on the air. At the least hint of dropping the series, he told Clark, "there is such a protest that I have felt that I must keep on."[39] In October 1929 the series was shifted to Saturday nights, a more lucrative piece of radio real estate, which Burgess regarded as a less favorable time for a sustaining program because he would have to compete harder to retain the slot. Even though the station wanted him to continue, he acknowledged that airtime had "come to have a considerable value in dollars and cents. One cannot expect a station to give for nothing time which can be sold at a good price."[40]

Once again Burgess tested the strength of his following by suggesting that, rather than changing to Saturday nights, it "might be just as well to drop the feature altogether." A "flood of letters" poured into the station—about thirteen hundred—listing three to four thousand names: "And such letters they were! They were from every walk in life. I did not dream of such a tribute."[41] All seemed well for a few weeks until the NBC network switched the popular *Amos 'n' Andy* show to 7:00 p.m. on Saturday—"which of course knocked me out," Burgess remarked. So, again, Burgess asked his Radio Nature League members whether they would rather have him appear in a fifteen-minute show at 7:00 p.m. or a half-hour show at 8:00 p.m. "I wish you could have seen the avalanche of mail," he wrote Clark, for it would "have settled any questions that may exist in your mind as to whether the general public is interested in nature.... We go on at eight o'clock."[42]

Ever the pragmatist, Burgess knew that this situation, too, was tenuous. Only one route might provide long-term stability: "If I could tie up with the right advertising concern I would be perfectly willing to do this, because it would establish the half-hour without danger of losing it and furthermore would give a chain hookup with the increased power that would mean."[43] In December Burgess auditioned for an NBC network program director in New York and had discussions with another chain about a coast-to-coast school network program, yet neither project ever bore fruit. CBS added Burgess as an occasional speaker on its "School of the Air" series, but NBC executives continued to regard Burgess as only a children's program host ("because I am a writer for children I must of necessity be talking to children") rather than a potential draw for more lucrative adult audiences.[44]

When Clark began in 1930 to broker another radio show, Burgess commiserated with the scientist on the "long slow process" necessary "to educate the powers that be to the fact that the public likes something besides jazz and more jazz." Nevertheless, Burgess remained convinced that an audience existed for "scientific matter... served up in a way that the layman can immediately grasp it."[45] Surely, science could be a worthy alternative to jazz.

SCIENCE IN THE AFTERNOON

In January 1929 E. E. Slosson suffered a major heart attack, confining him to home for many months. Despite this challenge, the Science Service staff had much to celebrate. Attention to science was continuing to increase throughout the general press. When Albert Einstein's paper on unitary field theory was published in Germany, interest was so great that a translation was cabled immediately and "appeared with all its mathematical formula on the front page of the New York Herald Tribune," something cited by Slosson as "conspicuous evidence" of the public's "eagerness" to "keep up with the progress of science."[46]

That January Science Service also began to consider accepting commercial sponsorship for its radio activities. A New York advertising firm developing a national radio series for Tidewater Oil Company had decided to devote fifteen minutes of a half-hour musical program to talks on scientific subjects by famous scientists. When it approached Science Service for assistance, James Stokley replied that the nonprofit would be happy to act as a broker and secure speakers for a flat fee of $250 plus expenses. He enclosed three pages of suggestions (from Nobel physicist Arthur Compton and astronomer Harlow Shapley to future U.S. vice president Henry A. Wallace, writer Paul de Kruif, and aviation pioneer Orville Wright).[47] When Slosson presented the Tidewater proposal to the executive committee, he said that no one on the staff "thought there would be any impropriety in Science Service appearing upon an advertising program like that of Tidewater Oil Co., provided that the compensation was sufficient and that the speakers [*sic*] would not have to insert any advertising matter into his talk and that the announcer made distinct that the talk was provided by Science Service."[48] Even though the advertising firm did not take up the proposal, Slosson and the trustees now seemed willing to entertain some type of commercial connection.

Around the same time, they began discussions with CBS about three other possible programs, initiating a relationship with the network that would last into the 1950s. The first proposal involved a national series of scientific talks to be broadcast through the CBS Washington affiliate, but this project was postponed by delays in establishing a permanent wire hookup between Washington and New York. A second overture involved a CBS series being planned with the National Education Association and intended for direct classroom use. The third possibility was to supply scripts of brief scientific talks (about two thousand words), called *Science Snapshots*, that would be included with CBS's Friday afternoon "Après Midi" hour. Here, finally, was an opportunity to produce material for national broadcast.

Everything fell in place quickly. Stokley sent the first script to New York by special delivery on April 3 for a CBS announcer to read on the April 5 show. The script was typical cut-and-paste. Watson Davis edited material already written for the *Daily Science News Bulletin*, smoothed the transitions,

FIGURE 8. James Stokley at WMAL radio microphone, circa 1926–1928. An astronomer and former high school teacher, Stokley worked as a Science Service journalist from 1925 to 1931 and participated in the organization's radio programs during that time, including as arranger and reader of *Science News of the Week* scripts. Courtesy of Smithsonian Institution Archives.

and added introductions and closings more suitable (or dramatic) for radio. The surviving scripts indicate that the program was aimed at an educated audience; sentences are unusually long for radio, and some technical terminology was retained.

The first segment on April 5, promising "the latest news on science," focused on astronomy and public health topics:

Mankind has always wondered what will be the final fate of the world about us. Answers have been given us by religious teachers and scientists. The other day Dr. Walter S. Adams, the director of the great Mt. Wilson Observatory in California, gave his ideas on the subject. Unless at some outlying parts of the universe, matter is being created, by some method of which we have no knowledge, the universe will eventually die, he thinks.... Dr. Adams is one of the world's most distinguished astronomers, and though his views are quite frankly speculative, his opinion is one that we must listen to with respect.[49]

The script then continued, with a few editorial changes, with the text of a *Daily Science News Bulletin* story called "Creation of Matter Needed to Prevent Death of Universe." In the second half, Davis brought listeners back to earth: "Whatever the ultimate fate of the universe, these changes will not take place in our time.... Of more immediate concern to us today are some of the diseases that man is subject to." He then described how, at the New York City Department of Health, "a young woman bacteriologist, Miss Georgia Cooper, has discovered no less than eleven types of pneumonia that have not been previously recognized." The script (drawn from a recent *Daily Mail Report*) discussed Cooper's work in connection with research on polyvalent serum being done by a female physician at Bellevue Hospital.

The *Science Snapshot* segment for April 26 opened with discussion of another female researcher, Florence Sabin at the Rockefeller Institute for Medical Research, praising her recent work on tuberculosis as "revolutionary."[50] Such attention to female scientists was unusual even for Science Service, which, by 1929, employed two women among its senior writers.[51] Although there is no evidence of this request in the surviving correspondence, CBS may well have encouraged the emphasis on women scientists because of the afternoon time slot and likely audience.

Other broadcasts followed similar rhetorical patterns, describing, for example, research on animal physiology. Stokley continued to try to sell CBS on a national series modeled on the local WMAL program ("I think that there certainly is as much popular interest in science as there is in politics," Stokley argued), but the network executive was not encouraging: "We are already having difficulty finding room for several well-established and highly popular programs. I am afraid that an evening's program would be entirely out of the question."[52] On September 17 CBS canceled use of the *Science Snapshots* scripts, explaining that the time slot had been "taken over by a commercial account" and there was "no other spot available for these talks at the present time."[53] The last segment aired on September 27, 1929. WABC (the local CBS station in New York City) also canceled its subscription to the *Science News of the Week* scripts. This was a tough blow—and worse ones followed.

On October 16 Slosson succumbed to coronary disease. Then, in the week after Slosson's memorial service, panic selling on Wall Street caused the stock market to crash. Science Service entered a long period of uncertainty, facing not only the economic challenges of the Great Depression but also a rancorous internal fight over who would control the organization.

DEPRESSION BLUES

After the cancellation of *Science Snapshots,* Science Service could have dropped its involvement in radio; the WMAL program and the script service had not been revenue-producing products. But staff and trustees alike now regarded broadcasting as furthering the organization's mission. In December 1929 Stokley attempted to broker another CBS series in a letter to William S. Paley, the head of the network: "Just a year ago I discussed with your Mr. Seebach the possibility of our presenting some scientific radio talks through the Columbia Broadcasting System."[54] CBS was establishing a new studio in Washington, perhaps they could develop a series of talks or a science news program? The overtures worked. By March 1930 Science Service was again on the air nationally, presenting weekly talks by prominent scientists through CBS on Friday afternoons, from 3:45 to 4:00 p.m. ("well-known scientists . . . are put on the chain here in Washington and after distribution go on the air from between thirty and fifty stations in a nation-wide hook-up reaching as far as San Francisco and Portland, Oregon").[55]

Given that the networks rather than local stations were now determining prime-time content, it seems extraordinary that the Science Service broadcasts aired at all. The programs lacked glamour and were dull competitors to the likes of Rudy Vallee, Kate Smith, and Eddie Cantor. They continued to follow the older presentation model, even featuring many of the same speakers as Austin Clark's Smithsonian series. By then people like C. G. Abbot (now the secretary of the Smithsonian), John C. Merriam (president of the Carnegie Institution of Washington), and Commander N. H. Heck (U.S. Coast and Geodetic Survey) were old hands at the radio microphone, delivering much the same talks on astronomy, paleontology, and seismology as they had for previous programs. Stokley's assurance to CBS of the quality of the speakers ("they can be depended upon for something of value") might have been written by Clark in 1923. University of Chicago professor Fay-Cooper Cole discussed "Race Problems as Seen by an Anthropologist" and

Harvard geologist Kirtley F. Mather talked about "Plumbing the Depths of the Earth."

Perhaps even more astonishing was Clark's brief reentry into radio. In January 1930 NBC asked if AAAS would cooperate on a national series of science talks. Clark, in his role as the director of the AAAS Press Service, represented the network's offer as "a high tribute to the regard in which the Association is held," but he also saw it as a chance to become personally involved again in broadcasting. He assured James McKeen Cattell that the speakers would be "men nationally known" whose information would be of "actual news value," and for the advisory board he enlisted scientists like Cattell, Robert A. Millikan, and John C. Merriam, along with "newspaper men of the highest standing."[56] Although these scientists lent their support, Clark soon found that duplicating the Smithsonian series would not be sufficient. By June the network was rescheduling programs, speakers were being pressured to discuss topics far beyond their texts, and the talks had attracted minimal press attention (Clark's initial justification for the project). In a letter to Cattell, Clark blamed the "recent somewhat sudden change in the public attitude toward science," which he said had "introduced a number of curious complexities in the selection and arrangement of radio talks, and in everything else having to do with bringing science to the people."

> Up to a year or so ago radio talks could be simple descriptions of facts in pure or applied science. Now scarcely anyone will listen to such talks, and they receive little or no press notice. The public seems to be satiated with the purely material aspects of science, and to be groping about in an effort to build a new mass philosophy.[57]

To some extent, Clark was correct in his assessment of what type of science might attract larger audiences. The local *Science News of the Week* show on WMAL had been moved to a more favorable time, 5:45 p.m. on Thursdays, but Science Service's national network show remained in the early afternoons (shifted in 1931 to 1:45 p.m.). As Stokley thanked one correspondent for his interest, he explained that scheduling was out of their hands. They were, he wrote,

> fully aware that an evening period for these talks would be much better than the Friday afternoon time.... However, this is a case where half a loaf is better than no bread at all. The times of all the broadcasters are tied up in the evening with commercial programs so that the only time we could secure for our talks, which bring in no income, was in the afternoon.[58]

Radio continued to have a whimsical side, especially at the local level. In January 1930 Stokley asked the WMAL station manager if he could "do anything to prevent" the orchestra from "noisy tuning and practicing during my talk." Such disturbances had happened for several weeks in a row, and listeners had complained to Stokley that they had difficulty hearing his words over the background noise:

> Of course, I realize that this orchestra is a paid program, and must tune and practice, but there is no use of us giving our time and energy, and the station giving the program time for our talks, if they cannot be understood.[59]

Stokley could handle such annoyances. The programs had been doing so well under his management that Maurice Holland asked Davis if the NRC's Science Advisory Council could hire Stokley half-time to supervise a limited series of popular science radio talks that the NRC was producing with NBC for dramatization at the Chicago Century of Progress ("to stimulate interest in science and its applications and to present over the radio the science philosophy").[60] The proposal may have been flattering but Stokley's first love was astronomy. On January 1, 1931, he returned to his hometown of Philadelphia to become head of the astronomical section and planetarium at the Franklin Institute and Museum, although he continued to write astronomy columns and features for Science Service. In addition to making most of the arrangements for the CBS talk series, Stokley had been the one who, with Frank Thone, read the *Science News of the Week* scripts on the local Washington station. Stokley's resignation transferred more duties to Davis, who now took on the lion's share of responsibility for arranging the radio series, in addition to his campaign to be Slosson's replacement.

A FIGHT FOR CONTROL

Slosson had been in fragile health throughout much of 1929. As a result, Davis had shouldered much of the burden of supervising staff, editing, and managing Science Service operations. When the chemist died, it might have been expected that Davis would be named acting (and, eventually, permanent) director. Instead, he remained managing editor, and Vernon Kellogg (a prominent scientist who was vice president of the Science Service trustees) assumed the title of "acting director." Although Davis had strong support from Ritter, the Scripps family, and the publishing executives on the board, a small group of trustees led by James McKeen Cattell was determined to replace Slosson with another scientist. Cattell minced no words, telling Davis, "You have doubtless known that we always intended to elect a scientific man of standing as head of the service, either as director, or as salaried chairman of the executive committee."[61]

The struggle over direction of Science Service emphasized persistent differences in how scientists and journalists perceived popularization.[62] To many scientists, the best popularizer was always another scientist; no journalist or other professional could ever be an adequate substitute. To those who embraced a vision of scientific popularization as free expression in a free society, the quality of the product mattered more than the writer's academic training. Was a story accurate? Did the product serve the audience's needs and satisfy its curiosity? Was the information useful?

Davis lobbied hard for the job of director. He had earned respect from many powerful scientists, yet his allegiances leaned unashamedly toward journalism and the public interest rather than embraced a blind defense of science. As chemist and trustee W. H. Howell wrote to Davis, "In spite of your protests I reckon you as a newspaper person, because invariably you take that point of view when debatable matters come up."[63] Scientists like Howell were convinced that selection of an eminent researcher to replace Slosson would preserve the organization's reputation among the scientific community. To the newspaper executives, however, a director who understood the marketplace would assure the organization's survival and growth. Davis pointed out to Howell and others that being a scientist did not guarantee access to newspaper offices. What mattered was whether "editors are confident of the authoritativeness and the reliability of our product."[64]

In early 1931 the executive committee offered the position to University of Missouri professor Winterton C. Curtis, a well-known zoologist who had little experience in either publishing or popularization. Curtis turned it down. Davis was then kept in limbo for another eleven months while Cattell and some of the trustees plotted to convert Science Service into a publicity machine for science. As Davis explained to J. W. Foster at Scripps-Howard Newspapers, Cattell and physicist Robert A. Millikan intended to abolish the policy "that we do not operate as a publicity organization, that we charge for everything we send out and pay for everything we get." Davis declared that "there will be a stiff fight on this score, so far as I am concerned."[65]

In the meantime, the economic depression deepened, and the newspaper business began to feel the effects. Science Service lost clients and income and was forced to cut back on the number of articles purchased from stringers. In October 1931 Thone and the rest began to write to prospective contributors:

It does not seem that we can buy any articles at present, as the economic situation has put considerable pressure on news publication. Prices paid for articles are less, and we do not often find it economically feasible to purchase articles from outside contributors.[66]

Then, the following January, they reluctantly informed even regular stringers that article purchases were temporarily suspended:

> It distresses me very greatly indeed to be compelled to inform you that because of the continued and increasing economic stringency, it has become necessary for Science Service to eliminate practically all purchase of outside material.[67]

They were, as Thone said, "having to batten down the hatches, shorten sail, and make ready to ride out the storm as best we may" because the economic recovery predicted for spring 1932 had not materialized.[68] Davis struggled to keep the main editorial group intact but had to reduce the office staff slightly and cut all expenses by over $10,000. As Davis wrote to Ritter, the budget cuts had made them

> somewhat shorthanded. It means more and harder work for all of us, although it may eventually prove to be a good thing for us to tighten our belts somewhat.[69]

Fortunately, subscriptions for *Science News-Letter* and syndicated products like *Star Maps* and *Science Shorts* remained steady, and the Scripps endowment money was intact. Davis reorganized the syndicated services, offering a "7-in-1 budget service" that allowed a newspaper to purchase more material for a single, reduced fee. The free radio script service continued with a list of over sixty stations, because it was regarded as a low-cost investment in promotion:

> Science Service receives no compensation [for the CBS talks]. . . . Neither does it pay for the radio time that is utilized by these talks. If we paid for the time used at commercial rates, it would involve an expenditure nearly as large as all expenditures made by our institution at the present time, over $100,000 annually.[70]

The organization also began to explore other popularization outlets as potential revenue sources. Electrical transcriptions of the radio programs were considered but were deemed prohibitively expensive.[71] In February 1932 Science Service produced a set of seven phonograph records for classroom use, each featuring a five-minute speech by a scientist (speakers included Karl T. Compton, Robert A. Millikan, and L. H. Baekeland) and an accompanying rotogravure portrait. Given the economic situation, sales of the records were not brisk, even at a modest price.

In April 1932, at the urging of Robert P. Scripps, a roundtable conference was held to reassess the Science Service mission and utility. Most of the speakers were scientists, and the criticisms and rejoinders echoed the debate raging behind the scenes. MIT president Karl T. Compton urged that the organization create "more effective scientific publicity."[72] The newspaper editors pled for "tolerance, patience and understanding" and encouraged both scientists and their colleagues in the press to abandon their "high-horse" attitudes. One editor remarked upon the arrogance with which scientists characteristically discussed the news business ("some scientists think they know all about newspaper work as well as their own").[73] This organization, he observed, had built its reputation by speaking authoritatively and accurately and avoiding "all suspicion of propaganda."

The stakes for civic education were high, Robert P. Scripps reminded participants:

> It is all very well to say scientific research will go on, whether any one outside the laboratories knows about it or not; or that the scientist can not take too much time off to consider the end result—the effect on humanity—of his labors. I challenge both positions.

> In the first place, the direct influence of the multitude cannot be underestimated. It is the man in the street, whom you have not reached, who is ultimately responsible for incidents like the Dayton, Tennessee, trial, and state laws which make such trials possible. It is the man in the street, and his attitude, that is ultimately responsible for appropriations to great state and national institutions, where much of the world's scientific work is carried on.[74]

MORE STORMS AHEAD

The speed limit in Washington then was a sedate twenty-two miles per hour, but the world outside the capital was changing rapidly, influencing science, science journalism, and the Science Service staff. A new administration assumed power in March 1933, and an Iowa friend of Thone's, Henry A. Wallace, became Franklin D. Roosevelt's secretary of agriculture. The spirit of celebration that usually energizes the town during inaugurations was subdued, however. The economic situation made everyone nervous. As Thone observed to one correspondent, "We feel very much like passengers on a ship during a dangerously violent storm: realizing the seriousness of the situation, we yet find it very interesting."[75] Even Roosevelt's radio chats, initiated during his first week in office, did little to calm the fears.

The letters that Thone and others were receiving from European scientists undoubtedly intensified their anxiety. One longtime stringer, Gabrielle Rabel, who had been working in a German laboratory, decided to return home to Austria that April; she told Thone that the world was just too "crazy" in Darmstadt.[76] On April 7 their stringer in Berlin, Theodor G. Ahrens, described the recent boycott of Jewish businesses and the rising anti-Semitism: "The National Socialist idea is: Germany for the German and the word German is supposed only to apply to those of Aryan descent."[77] Against this disturbing background, science was abuzz with news about the exhilarating advances being made in physics. In April, when visiting German scientist Otto Hahn lectured at the Carnegie Institute of Technology, he had discussed his work on "The Radioactive Elements and Their Use in Chemical Research." Other physicists were engaged in similar attempts to unlock the atom's secrets.

In May 1933, near the depth of the Great Depression, on the brink of one the most fascinating eras in the physical sciences, and while the European situation darkened, Davis was finally appointed director of Science Service. It happened rather unceremoniously. Trustee H. E. Howe wrote to Davis on May 24 that he did not even know of the appointment until reading a notice in the *Daily Mail Report*. Davis replied:

> The Directorship election came to me as a complete surprise and it was by mail ballot of the executive committee a few days after the annual meeting at the instigation of Dr. Cattell. We of the staff naturally feel quite pleased about this because we view it as a vote of confidence for the present organization. I wondered as you do why it was not done at the annual meeting but Dr. Cattell had evidently not made his decision at that time.[78]

The thirty-seven-year-old Davis took the reins at an extraordinary time. His entrepreneurial skills, pragmatic spirit, and liberal ideals helped the organization respond imaginatively and responsibly to the increasing connections between science and politics. In slightly more than a decade, members of the Science Service staff had become important nodes in a network linking scientists, government, industry, universities, press, and public. News of discoveries, the latest academic and political gossip, and information about career changes and controversies all came across their desks.

In 1933 they became a nexus for information of a different sort, and occasionally became involved in the events themselves, again crossing the line of journalistic objectivity. In late June another longtime stringer in Switzerland, physician Maxim Bing, wrote for help in securing a fellowship in the United States for cancer researcher Hans J. Fuchs, who wanted to leave

Berlin ("No official institution would dare these days to recommend a Jew. So the thing has to be done unofficially").[79] Davis responded immediately that he had sent copies of Bing's letter to various people in cancer research and at suitable foundations, expressing the hope "that for the sake of science as well as the individual a rescue of the potential victim can be effected."[80] Both Davis and Thone worked diligently over the next few years to assist scientists attempting to flee prewar Europe. One of their frequent allies was Stephen P. Duggan, whose bright young assistant at the Emergency Committee in Aid of Displaced German Scholars was Edward R. Murrow, with whom Davis would later work at CBS.[81]

Davis used his new power as director that December to "deputize" physicist I. I. Rabi as a special Science Service representative and arranged for a special pass so that Rabi could take a packet boat to meet James Franck, arriving from Germany on the *S.S. Bremen*, before Franck left quarantine and customs and might be grilled by reporters about his work. Davis also briefed Rabi on potential "safe" topics that Franck might discuss with the press, such as the high-voltage Van de Graaff generator at MIT or the Nobel Prize just awarded to Werner Heisenberg. And he added a final word of caution:

> May I also point out that it is quite probable that conversation between yourself and Prof. Franck on board the Bremen regarding the German political situation may be dangerous since I understand there are many ears open on board these ships today which lead directly to political channels.[82]

Finally, Davis added, if Rabi could then wire Science Service any information about Franck's visit that could be published, they should be glad to have it. The ruse worked as planned, and Rabi became one of the organization's long-term supporters.

During the time between Slosson's death and the permanent appointment of Davis, the radio efforts had continued, albeit modestly and with no illusions about the revenue potential. As Davis explained to one scientist, "We have two regular radio series both carried on as a part of our general work of the popularization of science, and neither of these series bring in any income whatever and both of them cost considerable in time and energy and occasionally direct financial expenditure."[83]

Although few final scripts survive from these years, the opening news segments undoubtedly reflected the political situation, making the same connections between science, consumer interests, industrial research, and government actions as did the print news reports. Radio programs in 1933 and 1934 discussed agriculture, new synthetic materials, and how science contributed to building skyscrapers, hydroelectric dams, and concrete roads. The guests exemplified an inclusive definition of "science"—industrialists, science entertainers, conservation advocates, biologists, physicists, and astronomers—ranging from noted chemists like W. Lee Lewis, who had created one of the world deadliest poison gases, to Sol Finkelstein, a Polish memory expert.

Davis and Thone exploited both professional connections and friendships in their search for news, topics, and guests. Washington was, in many ways, a small town. It was relatively easy to encounter potential news sources, to keep in touch with the scientific community at the Cosmos Club, and to become friendly with prominent politicians and government officials as well as new appointees. Davis had known J. Edgar Hoover, head of what would soon become the Federal Bureau of Investigation, for years; they had both grown up in the Capitol Hill neighborhood and were fellow alumni of George Washington University. Thone had a similarly extensive network of colleagues in biology, botany, and agriculture. When Thone's hometown friend Jay N. ("Ding") Darling became chief of the Bureau of Biological Survey in 1934, Science Service tendered its willingness to cooperate in the survey's work "for the restoration and rational use of our wild life resources" and Thone offered to sponsor Darling for membership in the Cosmos Club, just as he had sponsored Henry A. Wallace when the Iowan came to town.[84] Thone mentioned to one correspondent that FDR's right-hand man, Harry Hopkins, had even given him a lift home from a meeting the previous night. Such close ties to scientists, agricultural researchers, conservationists, political leaders, and the expanding worlds of engineering- and science-based industries were reflected in Science Service radio programs, just as they were exploited in its news reports.

BURGESS BACK ON THE AIR

In August 1930, after almost six years of broadcasting, Thornton Burgess canceled his series, in part because of time-consuming program production and correspondence but also because of lack of support from the station. "I presume that if I could set my talks to jazz music and sing them there would be no trouble in finding an advertiser," he wrote Clark.[85] He was "deluged" with complaints; some Radio Nature League members had even said that he "had no moral right to stop."

When Burgess returned three years later with a fifteen-minute sustaining program aimed at conservation, he regarded the series as an opportunity to demonstrate that an educational program could be entertaining and profit-

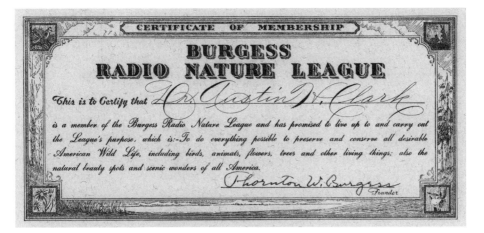

FIGURE 9. Certificate of membership in the Burgess Radio Nature League issued to Austin H. Clark and signed by Thornton W. Burgess, April 1935. Courtesy of David Pawson, National Museum of Natural History, Smithsonian Institution.

able. He began to search for a commercial sponsor to subsidize a staff so he could revive the Radio Nature League and "make it a factor in conservation." By early 1935, he had struck a deal with a Massachusetts chemical firm, Brewer & Company, Inc.: "Mr. Brewer, head of the concern, is a great outdoors man" and "eager to co-operate in any way to make the Radio Nature League all that it used to be and more." "I hated to go commercial," Burgess confessed to Clark, "but there was no other way out."[86]

By that time, Burgess was friends with many leaders in the widening national conservation movement, including Ding Darling, and he once again used the program to promote environmental values and wildlife preservation. "There has never been a time of greater need for active, enthusiastic work for conservation of our wild life and natural resources," he wrote to each member enrolled in the League.

There was a direct link between the warmth of the audience's reception, the sales of Mr. Brewer's products (primarily cod liver oil tablets), and Burgess's thirteen-week contracts. Members enrolled in the Burgess Radio Nature League received, along with their certificate, a personally addressed letter from the president of Brewer & Company and a "Dear Radio Neighbor" letter from Burgess that thanked the company for enabling the work of the League to continue and cheerfully commended "Sun-Glow Tablets" to "ward off colds and build up those inclined to be anemic."[87]

Between July 1935 and August 1936, Burgess used the company's sponsorship to publish a four-page newsletter, *Radio Nature League News*, distributed by retail druggists throughout the United States. In every issue, along with advertisements for Brewer pharmaceutical products, and wildlife photos and stories about animals or animal behavior submitted by listeners, was an editorial by Burgess. He frequently urged readers to lobby their congressional representatives to enact conservation legislation ("You need conservation, and goodness knows conservation needs you").[88] In September 1935 Burgess was still hoping to go on "one of the big networks before snow flies," but that dream was never realized.[89]

NEW DIRECTIONS

Science is not a gaudy parade with trumpets and flags, Davis wrote in his 1934 book, *The Advance of Science*. It is "a grand procession" that travels the "quieter roads of the intellect."[90] Through at least the end of that year, the Science Service radio efforts attempted to describe that procession with appropriate decorum. Each field of science received attention in turn: botanists and plant physiologists appeared on programs about the Morris Arboretum, drought, mosses, and Dutch elm disease; physicists and chemists came to the microphone to discuss "electron optics" and "friendly germs." The organization's reputation strengthened and respect for its radio projects grew. Davis found it easier to recruit guests, and, thanks to CBS's national network, the weekly "talks" could originate from seven different cities around the country, allowing Davis to take advantage of hot topics and hot researchers outside Washington, D.C.

The issue of who pays had been settled for Science Service's broadcasts by default. Although Davis continued to flirt with potential commercial sponsors, and the trustees remained open to the prospect, they were never able to make an acceptable arrangement with an advertiser. Their willingness to continue the series without such sponsorship reflects a commitment to broad-scale, mass popularization; their ability to continue demonstrates a willingness to provide content that conformed to the requirements of broadcasters and the interests of the listeners.

In that climate, juxtaposed against the uncertainty of the world situation, the excitement within physics, and few signs of real economic recovery, Science Service began toying with abandoning its traditional broadcast format and developing a program tied more closely to breaking news, centered on interviews, or involving dramatization. Each possible direction was influenced by existing trends within the radio industry. Each responded to indications of what audiences wanted. None of these paths were ones favored by academic scientists, educators, or the broadcast reform movement. They were, however, where the parade was heading.

—— ✳ ——

CHAPTER SIX

A Twist of the Dial

The audience walked out on us with a twist of the dial.
MAURICE HOLLAND, 1936[1]

AS the economic depression continued, attempts to use radio for adult education took on new significance, yet most of the major campaigns to accomplish that objective failed miserably. These projects also paid little attention to science. At a moment when science and engineering had special relevance to national recovery and were receiving increased attention in newspapers and popular magazines, the number of broadcasts aimed at educating and informing Americans about such topics remained small. Only a few individuals and organizations, acting independently of most educational reform groups, exploited radio for science popularization.

This failure seems all the more paradoxical because well-funded, well-connected national advocacy groups led by academics, including many scientists, and focusing on "radio in education" existed in the United States during the 1930s. The inability of these organizations to cooperate with each other, their intellectual snobbery and undisguised disdain for the very medium they were supposedly trying to utilize, and their unwillingness to invest significant resources in production of quality programs all hobbled their effectiveness. Moreover, these reformers seemed unconcerned with promoting science education for adults, choosing to concentrate on literature, music, economics, and public affairs (politics) rather than the ABC's of astronomy, biology, and chemistry. The physical and natural sciences were pushed to the sidelines in discussion and planning, and there is little evidence that scientific associations made much effort to push them back toward the center.

Nothing illustrates the emerging gulf between broadcasters and educators (and especially academic scientists) so well as the differences in how each group chose to gauge success in popularization efforts. To radio executives, the most important yardstick was how many people turned to a program "with a twist of the dial" and then stayed to listen. This number—and later, more sophisticated demographic measures such as age, residence, and potential buying power—determined how much could be charged for airtime. Even though the techniques of attracting listeners, holding them, and bringing them back week after week were becoming well developed, as were the techniques of advertising, the radio business remained centered on a sweet simplicity. The higher the numbers, the higher the profits. The higher the profits, the greater the success.

To scientists, the notion that "success" in broadcasting should be linked to popularity or the commercial marketplace was anathema. Scientific work is certified in print. Citation may be considered a measure of intellectual influence, but scientific accomplishment has never been judged by how many colleagues actually read a researcher's journal articles. Nobel Prizes reward creativity and ingenuity, not the number of papers published. In some fields during the 1930s, the number of people familiar with any individual scientist's work could be in the dozens, far fewer than the millions who read the science fiction novels of H. G. Wells. Discretion, dignity, and professionalism trumped fame, publicity, or applause. With the exception of Albert Einstein and Marie Curie, few modern scientists had become public celebrities. That era was yet to come.

When organizations like the Smithsonian, Science Service, and AAAS attempted to assess their radio ventures, they turned therefore not to the number of listeners or repeat listeners but to the reality they knew best: print publication. They published "official" versions of radio talks in scientific journals; they mimeographed and distributed copies of the scripts; and they produced supplementary "listener's booklets." Scientists clung to a communications medium they understood and believed that they could shape to their own standards, even though there were more egalitarian means available. By insisting upon measuring success through subsequent publication rather than a broadcast's direct impact and popularity, however, scientists were continually at odds with the radio industry. They were aiming at different targets.

Popularization remained a marginal activity for the scientific and academic elite during the early 1930s, rarely high on their professional agendas.

To some, popularization represented a necessary evil that, at best, drained resources and time from valuable research and, at worst, ran counter to science's best interests by subjecting it to the public spotlight. Neither attitude lent support to the struggle to gain airtime, an effort that required ever more ingenuity and accommodation to persuade listeners not to twist the dial away from science.

COMMITTEES AND CONFERENCES

The various uncoordinated campaigns to exploit radio for education found organizational expression during the 1930s in a number of government-funded studies and committees and in the formation of two major advocacy groups—the National Committee on Education by Radio and the National Advisory Council on Radio in Education. The agendas of these reformers were influenced considerably by contemporary intellectual critiques of popular culture.[2] As a result, their undisguised antagonism to mass-market radio, their naiveté, and their insistence on sober, unentertaining "education" made cooperation with the networks difficult. Their refusal to cooperate with each other further weakened their ability to broker change.

The National Committee on Education by Radio (NCER), which existed from 1931 to 1941, focused on radio's relationship to school-based learning and sought to redirect existing broadcasting resources toward education. Led by National Education Association official Joy Elmer Morgan, under-written by the Payne Fund, and endorsed by a coalition of social reform groups (including labor unions, the Paulist Fathers religious order, and the American Civil Liberties Union), NCER wanted to nourish strong, nationally based educational stations independent from the broadcast industry. Given radio's increasing commercialization, achieving this goal would have been difficult for any entity. Even if government pressure had been applied, many nonprofit institutions, struggling to survive in hard economic times, had begun to sell off their radio licenses to commercial broadcasters.

NCER's "cultural elitism" has been cited by many analysts as the foremost barrier to its success. The organization's leaders claimed to promote "formal and detailed [radio] courses on high culture, science, and government" and yet, Douglas Craig notes, that progressivism was shallow.[3] Historian Robert W. McChesney describes NCER rhetoric as having "a distinctly elitist tone," and Eugene Leach points to Joy Elmer Morgan's statements that America was in greater danger from "New Yorkism" than from "communism" as further evidence of NCER attitudes that alienated commercial broadcasters.[4]

Morgan's criticism could be scathing. He told the *New York Times* in 1932 that radio was "constantly searching for the lowest common denominator" and "tends to pull down rather than up." To Morgan, radio was just another agent of forces "bred in the hothouse metropolitan centers [that] will sap the ideals and the vision of the outlying regions which have been the stable centers of our national life."[5] Cooperation with broadcasters might invite co-option and imply endorsement of such depraved values.

The rival organization, the National Advisory Council on Radio in Education (NACRE), operated from 1930 to 1938 with funding from John D. Rockefeller Jr. and the Carnegie Corporation. NACRE focused more on adult education, and also encouraged cooperation between educators and the commercial networks and their fledgling organization, the National Association of Broadcasters. As a result, some historians dismiss the group as a "company union" for broadcasters. Nevertheless, NACRE holds particular interest for the history of science because its leadership included many prominent scientists and science popularizers, several of whom, such as physicist Robert A. Millikan, were associated with Science Service.[6] NACRE's national advisory board included MIT president Karl T. Compton, Bell Laboratories president Frank B. Jewett, and Columbia University professor Michael I. Pupin. Associations like the AAAS, American Chemical Society, American Psychological Association, and National Research Council were among NACRE's "cooperating organizations." NACRE's failure to emphasize science or engineering content is therefore all the more perplexing.

NACRE's director and guiding force was Levering Tyson, a professor at the Columbia University Teachers College who had previously been involved with another Carnegie-funded educational group, the American Association for Adult Education.[7] Tyson's public statements epitomized his group's accommodating approach to the commercial networks, such as when he declared that educational interests did not have any "inalienable right to the air" and must instead compete for the listeners' attention.[8] Privately, his attitude toward potential rivals was not welcoming and his characterizations of Joy Elmer Morgan and NCER were especially unflattering.[9]

Watson Davis became peripherally involved in NACRE activities as chairman of its Committee on Science. NACRE's Service Bureau also distributed copies of Science Service scripts. But Tyson apparently perceived all science popularizers as rivals, for he complained to Millikan that Davis "and others who are anxious to get prominent speakers on the air all the time are promoting their own interests to the exclusion of the larger aspects of the situation."[10] Tyson argued that NACRE should coordinate all requests for educational talks: "There is unlimited opportunity for good broadcasting on the part of qualified individuals, provided the Council can eventually be the recipient for all these requests."[11] Involvement by "befuddling" groups like Science Service and NCER would be dangerous to the overall objective.

These reformers confronted enormous challenges—how (or whether) radio would serve the educational needs of adult Americans, how to accomplish the practical tasks of producing and delivering programs, and who should pay. Tyson once observed that in these discussions educators and broadcasters frequently found themselves "on opposite sides of the fence, each group blaming the other for . . . short-sightedness."[12] Regrettably, little understanding emerged from the interactions, and the reformers' elitism and unrelenting criticism undermined their campaigns to increase educational content.[13] Both NCER and NACRE were also concerned more with education for education's sake, never clearly defining what "education" would mean or how a radio program's component subjects might be chosen other than for pedagogical tone and approach. Science played a nominal role in this vision for change. Only a few voices urged its inclusion.

A ROLE FOR SCIENCE

In addressing NACRE's first national assembly in 1931, Robert Millikan optimistically declared that radio "contains within it the possibility of exerting an influence toward introducing more rational, less emotional thinking and acting into American life."[14] He added that this power is accompanied by a responsibility "for avoiding looseness of statement, for getting the facts straight before presenting them." His remarks offered a typical appeal for using radio cautiously, seriously, and soberly—that is, for resisting the trend toward entertainment. But neither Millikan nor any other prominent scientist apparently proposed through NACRE any practical ideas for increasing the amount or quality of attention to science, and only the Committee on Psychology developed a substantive talk series.[15] NACRE, Douglas Craig notes, primarily tended to endorse broadcasts "that did not overburden listeners with intellectually demanding material."[16]

Even if NACRE and NCER had developed a creative array of programming, gaining airtime would have required educators to conform to broadcasters' requirements, something the reformers seemed unwilling to do despite broadcasters' initial receptiveness. CBS vice president Henry Adams Bellows explained to the first NACRE assembly that commercial stations did not automatically assume that educational programming would be beneficial to their interests. Content must attract listeners, because "all that a station has to sell is its ability to reach a listening audience."[17] On the other hand, Bellows emphasized, broadcasters know that audiences want more than transitory amusement. Entertaining and clever educational content would be welcomed.

Bellows's observations encapsulated the essential dilemma for science popularizers—they were caught between the sermons of the broadcast reform movement, which invested little in innovative program development but whose support and endorsement helped to legitimate popularization, and the practical requirements of the networks that controlled the platform and demanded a satisfied audience. A few, like Watson Davis, attempted to straddle the middle ground.

"Science," Davis noted when he spoke to the assembly in 1931, "is in the position of a kind, self-effacing, sacrificing, benevolent father to radio," having given this "instrument of pleasure and profit to the world, as it had done with other inventions such as the telephone, automobile, electric light, and so forth."[18] For the communication of popular science, radio's greatest effect will not be, he argued, "within the walls of schoolrooms" but in the "living-rooms and dining-rooms of the ordinary home with all ages in its audience." Like Thornton Burgess, Davis believed that radio would be most effective if the educational intent was disguised and if science engaged "all the mechanisms of oral and sound presentations that have been devised for stage, talkies, pulpit, rostrum, and radio." He admitted that this would not be easy—"many radio dials are shifted to music when the flow of entertainment is interrupted by a conventional science talk"—but he offered one suggestion:

> The least that we can do is to ease the listeners into the talks by theme songs. How effective it would be if Einstein, before his next radio talk, would render a violin selection and thus demonstrate to the radio audience the kinship of classical music and his revolutionary physics. One of Science Service's recent nationwide radio programs presented the very vibrations of the atoms themselves as musical chords.... In this case chemistry wrote its own theme song.[19]

"Science talks of the obvious sort," in which millions of listeners expressed interest, remained acceptable, but popularizers should also consider "all of the mechanisms used in advertising programs over the radio, in the talkies, in the broadcasting of news events over the radio." "It is just as necessary," Davis told the assembly, "to put showmanship into the preparation of science preparations as in the selling of toothpastes, coffee, and cigarettes by radio." Such statements would have been heretical to educational purists like Morgan and Tyson.

In fact, neither NCER nor NACRE appears to have had much concern for popularizing science. Each trumpeted the cause of education, but separate from the subjects of its lessons. While Scripps and Ritter had envisioned civic education in science as advancing the national good, these educators promoted civic education as a vehicle for moral reform. Despite elaborate promises and ample funding, NACRE brokered only fifteen brief series, primarily on topics in politics, economics, and psychology. Most of its assemblies emphasized regulatory issues or politics. At a 1935 conference cosponsored with the Institute for Education by Radio, Millikan suggested how to create "an intelligent electorate" through educational programs but failed to mention science as part of what that electorate should know.[20]

One of NACRE's few claimed successes, a thirty-two-week NBC series called *Psychology Today*, was the result of work by the American Psychological Association (APA). At Tyson's request, APA established a committee (Walter V. Bingham of the Personnel Research Federation, Paul S. Achilles of the Psychological Corporation, and Arthur I. Gates of Columbia University) that quickly and efficiently planned lectures on current research and pertinent applications of psychology. They surveyed APA members about effective methods of radio presentation, invited the most eminent psychologists in the country to appear (speakers included Floyd H. Allport, Arnold Gesell, Gardner Murphy, Edward L. Thorndike, and John B. Watson), and encouraged evaluation of the broadcasts by both listeners and APA members.[21] The psychologists set high standards for the presentations, which often combined technical information from the speaker's own research with attention to psychology's contribution to "understanding of the causes and control of human behavior."[22] The project also demonstrated the psychologists' expertise in understanding how humans learn. Lecture topics, Bingham wrote, should not be "abstruse or dull" or "leave the listener passive"—for only when a listener "is brought to do some thinking of his own will he really assimilate the facts and principles presented."[23] Each supplementary "listener's notebook" contained a section instructing the purchaser on how to get "the most out a radio lecture," with tips like "Listen, with pencil in hand, ready to jot down a word or phrase that will serve to recall the most important points [the speaker] makes."[24]

Commercial networks did produce some notable cultural programming in the 1930s, such as symphony broadcasts and debates on social issues, but there was little agreement about what actually constituted "radio education."[25] In 1935 the recently renamed Federal Communications Commission (FCC) assured Congress that commercial broadcasters were giving sufficient time and effort to educational programming and there was no need to set aside separate frequencies or reassign those being used for commercial purposes.[26]

PERSONAL HEALTH AND MEDICAL TREATMENT

After participating in the NACRE series, the American Psychological Association remained involved with radio during the 1930s in another role—attempting to persuade the Federal Radio Commission to investigate radio astrologers. Although protecting listeners from "pseudo-science" was rarely offered as a justification for the broadcasting activities of other scientific organizations, the proliferation of questionable "radio doctors" similarly pushed the American Medical Association (AMA) into action.

The most notorious of the 1920s pitchmen, John Romulus Brinkley, began aggravating the medical establishment when he acquired his own station, KFKB, in 1923.[27] On *Medical Question Box*, Brinkley diagnosed ailments based on the descriptions in patients' letters, gave medical advice to people he had never examined, and promoted, among other things, the use of goat prostate gland transplants to cure impotence. Eventually both of Brinkley's licenses (medical *and* broadcasting) were revoked, in 1929, but other dubious medical "experts" continued on the air. The AMA responded by producing its own broadcasts and by encouraging local medical groups and other credible health experts to go on the air to drown out the quacks.

The AMA began organizing its own monthly radio talks in 1923. These sustaining programs were expanded into a weekly CBS series in 1932, and subsequent series with NBC from 1934 through the 1950s. The AMA also encouraged, or at least did not criticize, accurate programs produced by commercial sponsors, although it adamantly rejected sponsorship for its own programs. On a show supported by Sharp and Dohme Pharmaceuticals, Howard W. Haggard (NBC-Blue, 1932–1933) discussed the history of medicine. Herman N. Bundeson's *Adventures in Health* was sponsored by Horlick Malt on NBC-Blue, 1932–1934, and American Cyanamid underwrote a medical interview show, *Doctors Talk It Over*, which ran on NBC from 1935 to 1936. Other popular radio physicians included D. R. Hodgon (CBS, 1928–1929), R. H. Baker (NBC-Blue, 1933–1934), Arthur J. Payne (MBS, 1935–1937), and Alan Roy Dafoe, the physician who had delivered the Dionne quintuplets (CBS, 1936–1938).[28] Physician Royal S. Copeland was on the air in various "health talk" series (1927–1932) while also serving in the U.S. Congress.[29]

Throughout the 1930s, various government agencies and nonprofit disciplinary groups, such as medical societies and public health organizations, promoted health education via radio. The New York Academy of Medicine pioneered in creating local broadcasts in New York City, and collaborated with CBS to produce *Highways of Health* from 1932 to 1943.[30] At least seventy-five state, city, or county medical agencies or associations regularly used radio during the early 1930s for public health education and announcements.[31] Dairy councils and commercial pharmaceutical and insurance firms produced other local and national medical series.

The most successful AMA series, *Your Health* (NBC, 1935–1940), was hosted by W. W. Bauer, director of the association's health education bureau. *Your Health* combined dramatizations with expert discussions, and continued into the late 1940s under various titles, offering a successful model for subsequent health projects. Contemporary critics sometimes complained that *Your Health* was "overdrawn," lacked sophisticated presentation, and featured dull and uninteresting guest experts, but the AMA remained convinced that the series offered valuable countervailing voices to "health racketeers."[32] Radio, Bauer argued, represented a huge "auditory billboard" that could "keep a certain message before the public" and thereby encourage them to be "health-conscious." To achieve positive results required using the latest techniques:

> Anyone who proposes to use the radio intelligently may as well begin by putting aside any Arabian Nights' fantasies about the breathless millions of the unseen audience hanging on his every word. . . . One who would be heard among the myriad voices of the ether must have something to say. But even that is not enough. He must say it in a manner which commands attention.[33]

MEASURING SUCCESS

In his NACRE presentation, Watson Davis had mentioned the "general problem" that continually bedeviled science popularizers—how to measure impact. When a physicist delivered a lecture in an auditorium, the audience's laughter, applause, or yawns provided immediate feedback. When a biologist published a book, the sales and reviews reflected public reaction. Scientists at first tried to evaluate the appeal of their radio programs through the content and numbers of personal letters from listeners. Gradually, they adopted other print-based measures.

When pressed to estimate the size of the audience for the Smithsonian show, Austin Clark initially claimed that his role was only to provide "first class and absolutely authentic" content and that broadcasters were "the ones primarily interested in the popular reception." He "merely acted as an agent in arranging talks for them."

> I have no definite idea as to how the Smithsonian talks on scientific subjects were received by the public, or any idea regarding the size of the audience. These matters concerned Station WRC from which the talks were given, and not me.[34]

This hands-off attitude had dissipated somewhat by Clark's second season on the air. He soon began calculating the numbers.

Quantitative measurements did not provide flattering pictures of most early science programs, of course, because audiences were relatively small. Reading fan letters offered a more satisfying experience. As historians like Susan Smulyan have described so well, during the early 1920s "long-distance listening" became a popular parlor game, stimulating informal competition among friends and "How Far Have You Heard?" contests sponsored by the stations. "Writing to radio stations became part of the fun of listening," Smulyan notes.[35] At institutions like the Smithsonian, where responding to public inquiries was part of the curators' routine, some effort was made to answer these letters and to tally their numbers. Within a few weeks of having stated that he "merely acted as an agent" and had no interest in audience measures, Clark wrote to another scientist that:

> The general public has become greatly interested in these talks, and from the last one given we received no less than thirty-three responses. This number indicates only the very few who took the trouble to express their interest in what was said—not one in a thousand of those who "listened in."[36]

He then added that the Smithsonian series had reached a stage "where it seems to be a marked success, with an estimated average audience of about two millions."[37] Where Clark obtained that unlikely number is not known—perhaps he was extrapolating from the local station's estimates—but he had clearly begun to search for some type of independent indicator of success. A few days later, he boasted to someone else that the three radio talks he delivered in the spring had stimulated as many as thirty-seven letters and many "requests for copies of the manuscript read, invitations to lecture, etc."[38] Clark asked a U.S. Department of Agriculture entomologist (who had delivered a radio talk about "clothes moths") how many letters or telephone calls he had received, adding, "The record for a single talk so far is sixty-five letters."[39] The Smithsonian's own annual report touted both the critical attention to the series from newspapers and magazines and the number of listener responses: "The series of talks was a distinct success, as attested by calls and letters from many listeners."[40]

Today, a few words muttered into a microphone (or captured on video) can survive to haunt a politician or celebrity for decades. In its early years, radio was an ephemeral medium—once you delivered your radio talk, nothing remained except a typewritten script and (perhaps) your ideas in the listeners' memories. Scientists accustomed to the self-satisfaction of seeing their names in print began to look for ways to capitalize on their broadcast efforts and to receive proper credit in a medium they respected. The scientists' preferred measure of quality rather quickly became not the number of letters praising a program or requesting more information but whether the talk was good enough to appear later in print. Within a relatively short time, typescripts prepared for delivery at a microphone were routinely published as articles in popular magazines, despite being based on well-known and already extensively published work. All radio talks, Clark declared, "should be written in such a fashion that they are suitable for subsequent publication as newspaper features or magazine articles, and also suitable for assembling in the form of pamphlets or small books which will meet with a ready sale."[41] Publication of the talks, he believed, insured "their permanent preservation."[42] During the AAAS annual meeting in January 1925, Clark brokered an agreement with James McKeen Cattell whereby manuscripts of the Smithsonian series would be offered for publication in the *Scientific Monthly*.[43] This arrangement continued through the life of the radio series. In June 1925 Cattell also began running the text of the series sponsored by the National Research Council and Science Service, and continued to reprint Science Service's radio talks through 1936.

In 1925 Clark and Cattell plotted a pamphlet series based on the Smithsonian talks. Clark attempted to persuade Harlow Shapley to add the Harvard Observatory talks to this project, but Shapley made his own deal for republication with the *Boston Herald* and *Harvard Alumni Bulletin*. In 1926 Shapley published the astronomy talks as a small book, sold for $2.00 a copy, and issued a revised version three years later.[44] The University of Pittsburgh academic departments published about half of that school's KDKA radio talks during the 1920s. There was an implicit assumption that listeners to science talks would want more explanation. As the Pittsburgh station manager wrote, talks could only "serve as beacons that point out paths of thought."[45]

By the mid-1930s, Science Service routinely repurposed its programs into magazine versions that could be sold or syndicated. Davis explained to medical professor R. G. Hoskins:

> In the case of a few of these radio talks we would like to use them as signed articles in our service to newspapers. When this is done we desire to send the speaker a small honorarium of $10.[46]

Hoskins replied that Cattell had already invited him to publish the talk in the *Scientific Monthly*; so Science Service opted for doing a "press release" in advance of the broadcast and a news story in the *Daily Mail Report*.

COUNTING POPULARITY

Once Science Service began to offer free mimeographed copies of scripts, Davis had a way to evaluate topics and guests in addition to the listeners' letters. Between June 1926 and November 1929 the staff recorded over two thousand requests for *Science News of the Week* scripts.[47] In 1934, when an MIT marketing and economics professor inquired about the measures Science Service used to gauge popularity, Davis responded:

> Free copies of the talks are offered in the concluding announcements of the broadcasts and during an average month some 200 requests are received. Topics in the fields of medicine, psychology and engineering are the most popular.[48]

The organization also got many nice letters from schoolteachers, he added.

Individual scientists were not immune to the flattery of public attention, and they too tracked such requests. Each time Henry Field of the Field Museum of Natural History appeared on a broadcast, he asked his staff to tabulate listeners' letters. A month after his March 19, 1935, talk ("The Story of Man"), Field told Davis that he had already received 250 requests for the scripts, from thirty-three states and two Canadian provinces.[49] After an appearance on February 23, 1939, describing "Recent Archeological Discoveries of the World," Field wrote that there had been 1,247 responses (230 letters and 1,017 postcards) from every state except Nebraska; Davis replied that, as of March 14, the Science Service office had received 1,420 additional responses to the same broadcast.[50]

It was a short step to combining the texts and other material into some sort of supplementary bulletin and tabulating the sales. The APA's *Listener's Notebook*, produced with the American Library Association, included explanatory illustrations, bibliography, discussion questions, and a "report form"

U. S. OFFICE OF EDUCATION • FEDERAL SECURITY AGENCY
SMITHSONIAN INSTITUTION RADIO PROGRAM

NATIONAL BROADCASTING COMPANY

The
WORLD IS YOURS

March of Science

Published by COLUMBIA UNIVERSITY PRESS
Vol. 1, No. 15 • Jan. 15, 1940 • $2.00 a volume, 10c a copy

FIGURE 10. Cover of the March 1940 educational supplement for *The World Is Yours*. These booklets accompanied the Smithsonian Institution radio series and featured script excerpts and articles by Smithsonian curators and scientists. They are typical of the printed booklets sold to supplement radio programs. Courtesy of Smithsonian Institution Archives.

that listeners could mail back to NACRE with their opinions on the subjects and quality of the broadcasts. None of the NACRE materials was distributed for free, however, despite the council's comfortable underwriting. For a series on economics ("Economic Aspects of the Depression"), reading guides sold for a dollar each; listener's notebooks for the six psychology programs sold for twenty-five cents and lecture reprints for an additional fifty cents.[51] By May 1932 more than forty-five thousand of the psychology series notebooks had been sold, and a "snowstorm" of mail and telegrams had arrived from listeners in cities and on farms, from Canada to Hawaii.[52]

Other groups combined script excerpts with short essays or reading lists into booklets offered for sale. To accompany the Smithsonian Institution's *The World Is Yours* (see chapter 7), the U.S. Department of Interior's Office of Education created a magazine eventually published and sold by Columbia University Press. The *World Is Yours Magazine* contained instructive illustrations like charts or maps, suggested reading lists, script excerpts, and a fifteen-hundred- to two-thousand-word summary of each week's topic by a Smithsonian curator or scientist. The first issue, January 1937, contained ethnologist Matthew Stirling's article on Mayan Indians and engineering curator Frank A. Taylor's discussion of the evolution of mechanical timepieces. For the first three years of the project, the federal agency distributed the magazine for free but then decided that readers should share the cost. Beginning in October 1939 it charged ten cents per issue. On occasion, the businesses or organizations featured in *The World Is Yours* scripts would purchase copies of the magazine for promotional use. The Portland Cement Company, for example, purchased a thousand copies of an issue featuring a broadcast titled "The Story of Portland Cement." By June 1940 the project celebrated the sale of the hundred thousandth copy of the magazine, but production was canceled a year later. Federal underwriting had run out, and the Smithsonian was unwilling to subsidize publication.

Simply counting the number of script requests or booklet sales did not, of course, provide an adequate measure of "public understanding." How did people actually use the information they gained from listening? Did a broadcast motivate them to learn more about the topic or science in general? None of the scientific associations seemed interested in asking such questions. For its 1938 series *Science on the March*, AAAS mailed, upon request, 170,600 free copies of the printed scripts, then used that same list to survey audience response. Apart from full-time students (about one-third of the respondents), the listeners were primarily professional, white-collar workers and teachers, in proportions higher than in the American workforce overall, and most had listened with family, friends, or schoolmates and had discussed the program with them.[53] Almost 28 percent of the respondents said that they intended to adapt the scripts for dramatic readings at home, school, or social club, but AAAS gathered no further data on the use of what they had heard.

Perhaps acquiring further data did not matter. The assumption that radio's "social value" should be calculated on the basis of its educational effects ignored the implicit social transactions taking place at the moment of broadcast.[54] Americans purchased their own receivers but did not have to pay to listen to radio. Instead, they gave something considerably more precious—their time and attention to the programs and commercial advertisements. Broadcasters added to the transaction when they invested in production facilities, transmitters, and talent, and donated airtime for public service programming. Stations were happy to have their off-peak hours filled with reliable content. In return, the public service programming was expected to attract listeners to the commercially sponsored programs.

In comparison, the investments being made by scientists and scientific organizations during the 1930s were relatively small. Programs that consisted of roundtable discussions among experts, or fifteen-minute talks by scientists on topics they had lectured about many times before, cost the scientists and their professional associations very little, even though the potential social return was considerable. As Clark acknowledged, radio enabled the scientific organizations "to reach a larger section of the public" and thereby benefit from "a large amount of valuable advertising which, though received without cost to them, means a considerable expense to others, who are quite justified in anticipating that this expense shall in some way be counterbalanced."[55] Given the relatively primitive state of audience measurement techniques in the mid-1930s, and given the scientific community's unwillingness to gauge success by marketplace popularity, scientists and their organizations could not be sure whether their programs had had any impact at all.

POPULARIZATION AS A "NECESSARY EVIL"

The uncertainty over how to measure the success of popular science also reflected persistent disagreements over its purposes. Some leading scientists, like MIT president Karl T. Compton, regarded popularization as "publicity" for science and thought that it was justified as a means of advancing public education: "All of us who believe in education as the basis of satisfaction in living and of a democratic government look upon educational work as a high public opportunity and responsibility."[56] From this perspective, popular science possessed high social value and should be regarded as a professional responsibility, part of every intellectual's obligation to share knowledge with society.

Others, however, characterized popularization as a "necessary evil" linked to cultivating a positive public image or to campaigns for political support and funding. Austin Clark synthesized many such ideas in essays published during the early 1930s, and those writings offer useful windows on the prevailing sentiments within the mainstream scientific community. Science educators like Benjamin C. Gruenberg, for example, argued that popular science, in all its forms, indulged too much in "extravagant emphasis upon the spectacular" and had begun to reduce science "to a modern form of magic."[57] If Science Service represented the "democratic" face of popularization, choosing content according to audience needs and interest, then

Clark and Gruenberg articulated the opposite view: Experts, not the vox populi, should select whatever popularizers broadcast.

The elitists did acknowledge that radio afforded an expedient route to reaching large numbers of people and therefore to reinforcing public support. In "Selling Entomology" (1931), Clark admitted that "science is a business." "In order to prosper," he wrote, "it must be advertised and sold, much like any other business, and in much the same way as any other business."[58] Building a positive public image was thus in the self-interest of universities, scientific associations, and science as a whole. Radio offered a tool with which to shape that image. When Smithsonian staff member W. F. Austin attempted around the same time to persuade Secretary C. G. Abbot to consider working with NBC on another series, he spelled out the value for the institution in blunt terms:

> Our interest is in helping to make the general public science-minded. The Institution's interest should be to make itself better known to people. A public sentiment can be built up which you can utilize in any endowment plans you may work out in the future.[59]

Clark believed that radio could transform American culture: "The power of the radio in promoting conversation and for the scientific education of our people cannot be overestimated," he told Burgess. "It reaches scores of thousands who seldom or never read the printed page but whose thirst for accurate information is insatiable."[60]

The dilemma facing these scientists was that respect for formal education seemed to be declining just as the power of radio and other forms of mass culture were in the ascendancy. At one time, education inculcated "a great respect for knowledge." Now, as Clark spelled out in "Science and the Radio" (1932), schools encouraged the presentation of information "as nearly as possible in the form of a pleasant amusement, with an underlying idea of its potential economic value rather than of its cultural significance." The public, he added, does not turn to radio for education. Instead, audiences seek "amusement or diversion." To reach people accustomed to these entertainment approaches, scientists must apply the same principles exploited by the newspaper science writers who compete daily for attention with "murders, scandals and similar and comparable phenomena."[61]

This was a radical suggestion. In an early draft of this essay, Clark had actually suggested that Americans, while interested in "new facts and new ideas," were intrinsically suspicious of being "educated." Success, then, would require deception: "Radio talks given in a manner even remotely suggestive of a desire to instruct are bound to be pathetic failures."[62] Public education must maintain a delicate balance between what is good for science and what is good for the audience. Above all, Clark believed, scientists must not relax their own high standards or expectations for accuracy, completeness, and dignity.

> If the people want education in science they will seek it; also they will be willing to pay the bills as they are now doing for the science they get in the newspapers and magazines. If science is sugar coated, or immersed in mental alcohol, and presented to them on a golden platter, or in a golden goblet, they will lose respect for it.[63]

The listening audience determines what stays on the air, he explained—either through their expressed preferences or through broadcasters' assumptions about what listeners will like. Clark called the processes of attending to those preferences the "grim realities" inherent in radio broadcasting. Success depended "upon the willingness of the audience to continue to pay the bills." Moreover, he wrote in "Science and the Radio," scientists must dismiss the belief that general audiences either "wish to be like us, or wish to know as much as we do or are to any appreciable degree interested in what seems all-important to us." Popularization was a necessary evil, an expedient to a greater good.

Although more discussion of popularization began to appear in scientific journals during the 1930s, resistance within the scientific establishment did not disappear. Physicist Joseph Sweetman Ames, president of Johns Hopkins University, for example, declared in 1934 that he did not even "approve" of efforts by respected organizations like Science Service, Associated Press, or the *New York Times* "to present to the public the results of a great deal of scientific work." "Almost without exception," he wrote, they emphasize the wrong aspect of the science and sensationalize modest advances—"I deplore the attempt to make 'news-worthy' [the] contributions from scientific laboratories."[64]

"FRANKENSTEIN OF THE WINGED WORD"

Maurice Holland, director of the National Research Council's Division of Engineering and Industrial Research, looked at popularization quite differently. He minced no words in blaming the scientific community for failing to realize radio's potential for presenting science. After a brief flirtation with the medium, he said, "science has abdicated from the throne of power."

> Science which created this Frankenstein of the winged word has gone back to its first love—the love of science for science's sake.... The voice of science that gave the infant prodigy to the world is still. Science is the stepchild on the doorstep of radio.[65]

Science was not absent altogether from radio, of course. As Holland pointed out, its name and authority were being invoked continually within commercial announcements:

> The name of science is frequently called upon as authority to uphold the virtues of many products. Everything from a toothpaste to a face cream is being extolled on the radio in scientific exaggerations that would make even the movies blush.[66]

Broadcasters had played a role in creating this situation. The owners of the networks, Holland acknowledged, know that they "must give the public a show if they are to stay in business." But the scientific organizations' refusal to acknowledge this commercial basis—that is, to admit that the audience "likes a show" and "wants to be entertained"—placed scientists at a disadvantage in any negotiations for time. Moreover, to date, most attempts to dramatize science had been "mediocre." As a result, Holland explained, valuable opportunities had been lost:

> We were given the ear of Mr. and Mrs. America, and then some pseudo-scientists and spotlight-grabbers became the "Voice of Science." They are not doing very much good, either to science, or to radio, or to our chance with the public.[67]

The "audience walked out on us with a twist of the dial."

What, then, could be done? A decade earlier Holland had urged the NAS and NRC to remain actively engaged in radio popularization, and he remained optimistic that scientists could reclaim the spotlight and attempt to tell their own story.

> Fifty million people in the radio audience are waiting to hear fifty thousand scientists tell their story.... Are [the scientists] willing to co-operate with the showmen in dramatizing the romance, the success stories or even the mysteries of science for the radio audience?[68]

For those who regarded popularization as a necessary evil or who, like Joseph Sweetman Ames, dismissed it altogether, Holland offered unwelcome advice. Eventually, in the mid-1930s, groups like AAAS, Science Service, and the Smithsonian did begin slowly to test the boundaries of acceptability by incorporating interviews and dramatizations within their broadcasts. Dramatized science responded to the fifty million Americans who had been waiting for scientists to "tell their story." The challenge to popularizers was how to dramatize those stories without eroding science's cultural authority, and without betraying its hard-won reputation for accuracy.

CHAPTER SEVEN

———— ✳ ————

Facts and Fictionalizations

Actors move in and out . . . of these diversified hours with syringes, folios, bottles of snake venom, ether cones, bird notes, and Irish poetry. Every once in a while they smash an atom or put someone to sleep in a new way.

MAX WYLIE, DESCRIBING *THE HUMAN ADVENTURE*[1]

RADIO programs whose titles promised "new horizons," "adventures," "explorations," and "science on the march" harmonized well with 1930s political rhetoric and the New Deal ethos.[2] Science popularization too was infected by this spirit of social optimism. Eager to seem dynamic and forward-thinking, many of the creators of science series began to adopt dramatizations and similar entertainment techniques in the service of what Smithsonian secretary C. G. Abbot dubbed "ultra-popularization."

In 1931 Watson Davis had somewhat prematurely touted dramatization as one of "the most effective means of presenting science over the radio." "Radio dramas of history in the making," such as the dedication of telescopes, he declared, could help the public "to appreciate the drama, the romance, and historical perspective of science."[3] Four years later, prompted by Harry L. Smithton and James McKeen Cattell, the Science Service trustees debated the possibility of dramatizing the Science Service talks. Perhaps the talks "should be put in dialog form," Smithton suggested.[4] It was an odd discussion, tucked amid attention to more serious financial and administrative matters, and hinted at the scientists' willingness to consider using entertainment to achieve civic education.

Something was needed to help science rise above the noise, to increase the amount of science content available. The rest of radio was dominated by comedies, dramas, and music, and in May 1935 a Radio Institute of Audible Arts brochure listed just one nationally broadcast science program, CBS's *Science Service Talks*. That situation soon began to change, thanks to a number of corporate and government radio initiatives that experimented with "dramatizing" science, adding dialogue and fictionalization to accounts of great experiments, employing professional actors to read the words of revered scientists, and using orchestral accompaniment and sound effects to evoke dramatic moods. The most notable projects—DuPont's *Cavalcade of America* and the Smithsonian Institution's *The World Is Yours* (produced with the U.S. Office of Education)—relied on substantial corporate and government underwriting. Others, such as the American Museum of Natural History's *New Horizons*, combined private initiative with network production. The more listeners were drawn to such series, the more other programs imitated their approaches.

The adoption of entertainment techniques for educational goals revived the unresolved questions of whose intellectual standards should prevail. The dramas sometimes teetered between acceptable creative license and embarrassingly inaccurate fictionalizations. Moreover, the methods that dramatists used routinely—narration to set the scene, music to set the mood, evocative sound effects, bits of extraneous dialogue to establish a character's personality—left less time for thorough (or sometimes any) explanations of the science involved. By comparison, comprehensive technical discussions seemed irretrievably dull. During the mid-1930s, the incorporation of ever more dramatization shifted the emphasis in science series away from facts, away from presenting science as a body of knowledge or as a systematic approach to understanding. Popularization was edging toward its current focus on personal, political, and moral aspects of the scientific enterprise and toward discussion of applied science rather than basic research.

Although intertwining fact and fiction helped to attract new listeners to science programs, this approach demanded a higher level of audience attention and content discrimination. How could listeners tell whether the assured and imperious voice they heard was that of an actual scientist or an actor? How could listeners with little or no scientific training distinguish between fictionalized fact and outright fiction if both were presented in authoritative and seemingly technical language? The implications of this changing context for popularization became unmistakable in October 1939, when Orson Welles's radio adaptation of *The War of the Worlds* triggered panic among inattentive listeners. An evening of pre-Halloween fun epitomized the swirling fact-with-fiction environment in which science was being broadcast and being received.

COMMERCIAL CAVALCADES

When large science and engineering corporations first underwrote radio programs, they preferred straightforward expert talks or similarly serious presentations. One of the longer-lasting series, General Electric Research Laboratories' *Science Forum* (later called *Excursions in Science*) was broadcast from 1936 to 1949 over the company's radio station, WGY, in Schenectady, New York, and via two international shortwave stations. Although *Science Forum* did not explicitly mention its commercial products, General Electric wanted the programs to emphasize "the achievements of modern research and engineering" and to encourage "appreciation of scientific endeavor."[5] Each program opened with a ten- to fifteen-minute talk by a scientist, engineer, or college science teacher, followed by an interview, panel discussion, or "Science Exchange" segment during which experts answered general questions on the topic. *Science Forum* audiences responded enthusiastically to the question and answer segment. In the first three years of "Science Exchange," listeners submitted over six thousand questions on topics ranging from relativity theory to what makes water freeze.[6]

Through the years, guests on the General Electric programs included such prominent scientists as chemist Irving Langmuir, physicist Hans Bethe, and microbiologist Selman A. Waksman. Researchers discussed "adventures within the atoms" and "chasing the Moon's shadow," and described heat pumps, fireflies, bats, dodos, phosphors, sunspots, and meteorites. Each guest's material was rewritten by a skilled copywriter, and a character (called the "reporter in science") set the stage for the scientist's remarks. Later, in the 1940s, the series began to include cautious dramatizations of accomplishment and invention, often introduced by a real scientist or engineer.

The most lavish corporate-funded radio drama in the 1930s was *Cavalcade of America*, underwritten and sponsored by E. I. duPont de Nemours & Company (DuPont). Part of a campaign devised by Bruce Barton and his advertising firm Batten, Barton, Durstine & Osborn (BBDO) to redeem DuPont's reputation and corporate image from its association with weaponry, the project also generated the famous slogan "Better Things for Better Living... through Chemistry." *Cavalcade of America* debuted in October 1935 with an inspirational drama about the *Mayflower* pilgrims. From 1935 to 1952, DuPont funded 791 radio episodes and then turned to television, sponsoring

FIGURE 11. DuPont Company exhibit advertising *Cavalcade of America*, December 1938. This holiday display in an Atlantic City, New Jersey, storefront illustrates how the corporation saw the radio programs as valuable promotion for its "Better Things for Better Living through Chemistry" campaign. The poster at left trumpets historian James Truslow Adams as an expert consultant to the series. Courtesy of Hagley Museum and Library.

another 197 tele-dramas from 1952 to 1957. Because the program proved to be popular with both audiences and critics, the substantial corporate investment was considered worthwhile.[7]

Not every *Cavalcade* episode featured science or scientists, but each program closed with a brief reference to the "story of chemistry," linked in some way to DuPont products, and the series glorified individual achievement and promoted an idealistic view of social progress. Scripts used historical events and people as analogies for the corporation's own goals and approaches, declaring, "The research chemists of DuPont laboratories are carrying on in the same spirit, discovering and creating useful products so that the American people may have better things for better living."[8] When a program had a science or medicine theme, the introduction often emphasized that the goal was to inspire, not necessarily to educate. In the ninth episode ("Heroism in Medical Science"), for example, the narrator explained that the story would exemplify heroism and the "faith and courage" of "American pioneers."

> Where primitive man feared the power of Nature, civilized man utilized that same power to his advantage. In our own country more than in any other, Nature has been harnessed to improve our living conditions. We defy the elements and refuse to let them conquer us.[9]

A 1936 episode called "Pioneer John Winthrop of Chemical Science" informed listeners that Winthrop's life would give them "new understanding of the tremendous handicaps under which the pioneer chemist worked," as well as a "better appreciation" of how chemists in the 1930s were "striving to create a new and better America."[10]

Although BBDO hired noted historians to advise on topics and review scripts, the writers had carte blanche to fictionalize in order to entertain. The Winthrop episode opened with the colonist experimenting at home with using potashes to make soap. "'Tis hard to keep the house clean, John, with all those kettles and pots lying about," Elizabeth Winthrop complains. When a neighbor criticizes Winthrop for spending time "on this foolishness" when there is land to be cleared, he responds: "There's a place for anything that fills a need—and our needs are great. Some of them can only be met through the science of chemistry."[11] In another episode, the writers contrived an accidental meeting between scientists Robert Hare and Benjamin Silliman, in which Silliman declares (in suspiciously inspirational language) that "the progress of science is like an endless chain, Mr. Hare, each link joining what is behind and what goes before."[12]

Through the years, the DuPont project employed many accomplished writers, actors, and directors. Erik Barnouw wrote a script about research on Rocky Mountain spotted fever, adapted, with permission, from Paul de Kruif's book *Men against Death*. Sinclair Lewis was paid to adapt his own novel *Arrowsmith* for an episode starring Tyrone Power and Lurene Tuttle. In 1941 there was an adaptation of the Pulitzer Prize–winning medical drama *Men in White*. Episodes routinely included Hollywood stars like Edward G. Robinson (as neurosurgeon Harvey Cushing), George Sanders (as Benjamin Rush), Douglas Fairbanks (as Benjamin Waterhouse), and Lionel Barrymore (as Luther Burbank).

By the late 1930s and early 1940s, other companies had followed DuPont's example of blending inspirational drama with science and technology themes.

FIGURE 12. Live broadcast of *Cavalcade of America*, New Brunswick, New Jersey, April 19, 1948. The episode, titled "Winner Takes Life," celebrated the career of Rutgers University microbiologist Selman Waksman and his discovery of streptomycin. The episode starred Paul Lukas, Ann Rutherford, and Jackie Cooper. Courtesy of Hagley Museum and Library.

The Westinghouse Corporation, for one, began sponsoring *Adventures in Research*. Its modest dramatizations, introduced by a Westinghouse scientist or engineer, explored such topics as development of an efficient internal combustion engine ("The Paper Invention That Worked—Rudolf Diesel"), astronomy, synthetic rubber, X-rays, and even poison gas warfare.[13]

INVESTING IN SHOWMANSHIP

In the years following cancellation of Austin Clark's program, the Smithsonian had cooperated occasionally with other radio projects or had supplied speakers for broadcasts but had not attempted production of another series. Money was always the issue. The institution balked at investing in formats like radio, preferring to focus on print publications, lectures, and exhibitions. Then the cultural and arts component of the New Deal provided an opportunity to tap federal funding. The result was an innovative drama series called *The World Is Yours*.

After the first meetings in spring 1936, Federal Security Agency official William Dow Boutwell worked diligently to broker an arrangement among NBC, the Smithsonian, and the U.S. Department of Interior's Office of Education. Within a few months, scripts had been written and approved and programs were on the air.

The Office of Education was already producing other network radio shows, including *Have You Heard?*, an "everyday science" series that simulated "dinner-party conversation about scientific facts" with plots and topics inspired by current events.[14] The master of ceremonies for that show, a character called "the Friendly Guide," sought to dispel misbeliefs and superstitions and to inspire "love for nature." Such projects had been considered experiments in determining "the type of educational programs that would both instruct and entertain," and the FSA now wanted to develop a dramatized series.[15] The Smithsonian offered an ideal partner for another radio experiment—a great and prestigious cultural and scientific organization with enormous collections containing endless ideas for programs. It also employed hundreds of experts, who would be available for consultation and on-air interviews.

The opportunity was also favorable for the Smithsonian, which was continually searching for ways to enhance its public reputation. As Boutwell emphasized to Abbot, the program "would help make known to millions of people the work and contributions of the Smithsonian Institution."[16] By early May, Boutwell and the U.S. commissioner of education, John W. Studebaker, had convinced NBC to set aside time for a nationally distributed series to premier June 7. The economics of the arrangement favored the network and the Smithsonian. American taxpayers, through the Office of Education, underwrote all production costs, including professional scriptwriters, and subsidized publication of the supplementary booklets and a monthly magazine whose circulation grew eventually to 150,000.

Critical to NBC's involvement was the authenticity and authority brought by the Smithsonian connection and the potential of a public service program that might be entertaining and also attract new listeners. For the Office of Education and the Smithsonian, the paramount concerns revolved around accuracy. Other Office of Education radio programs had appointed committees of experts to review all scripts, including "from the point of view of policy." A similar arrangement was proposed for this program: "Each script must be made completely satisfactory to the Smithsonian Institution before it goes on the air," Studebaker promised Secretary Abbot.[17]

Instead of appointing a committee, Abbot designated Webster P. True as the official liaison. True had considerable political influence within the institution, primarily through his role in controlling the quality of all print communications, including the Smithsonian's prestigious Scientific Series. True began mediating between NBC, Studebaker's office, and the experts, adjudicating any disputes over content. In consultation with True, the curators and scientists suggested and selected topics; then Office of Education staff would consult with each expert, draft a script, confirm its accuracy with True and the expert, and transmit it to NBC for further creative reworking. Rather than being entrepreneurs, on-air voices, or decision makers, the Smithsonian scientists stood on tap, along the edges. And unlike Clark, True never stepped to the microphone himself; his role remained behind the scenes, as impresario and quality control officer, guarding his employer's interests. True also worked closely with Office of Education staff on all publicity, and he approved any supplementary material distributed to listeners.

The use of drama was not without controversy, especially for a government-funded venture. Some agencies, like the U.S. Department of Agriculture, were already experimenting with dramatic techniques, at first using their own employees and then, at the urging of the networks, employing professional actors. This practice, Jeanette Sayre noted in 1941, prompted considerable debate about the merits of "documentary" versus "dramatic" techniques. Many critics argued that dramatization introduced elements unbecoming to government-sponsored messages, especially if the program's topic touched on controversial policies.[18]

The Smithsonian project began with thirteen programs intended "to show the person who can't come to Washington what he would see and feel if he visited" the museums, starting rather naturally with an episode describing the institution's founding and history.[19] Aside from another predictable choice ("Fashion Parade of Presidents' Wives," which described a famous exhibit of inaugural gowns), most programs during the first season focused on scientific or technical topics—astronomy, zoology, anthropology, and aeronautics. From the outset, there was a deliberate attempt to connect basic science to the improvement of life, to emphasize that while something may be the pinnacle of erudite learning, it may also be useful, as indicated by such titles as "Exploring the Ends of the Earth with a Definite Purpose."

The World Is Yours scripts embraced a type of dramatization seemingly incompatible with the institution's dignified reputation. The topics (chosen by the experts) signaled an appropriate sense of seriousness, while the dramatizations themselves trivialized that content. By the second and third years, the series balanced about one-third cultural and art history topics (e.g., Abraham Lincoln artifacts and legends) with scientific and technical topics, such as discussions of whales ("Largest of Mammals"), snakes ("Big and Little"), electricity, rockets, and forest conservation. The format and topics proved popular. After the first year, *The World Is Yours* was rescheduled from its original Sunday morning slot to a more favorable time on Sunday afternoon.

One continuing feature was a fictional narrator called "Oldtimer" (the "lovable mentor of the Smithsonian"), whose role was to interact with other characters, such as his grandson "Skipper," and to introduce the vignettes, as in this episode from an October 1939 program about Portland cement entitled "Man-Made Stone":

> *Mr. and Mrs. Shalit, while on a tour of the country, were impressed by the extensive use of concrete in the United States. When they visit the Smithsonian, Oldtimer invites them into his office for the story of the most important ingredient of concrete—Portland cement.*
>
> MRS. SHALIT: Go on, Oldtimer—where did Portland cement get its name?
>
> OLDTIMER: Well, now—we must set our imagination-clocks back about 128 years—that'll bring us to 1811. Napoleon is Number One Man on the European Continent. Beethoven is hard at work on a new symphony. In England... Joseph Aspdin, a bricklayer is puttering about the kitchen of his tiny home....
>
> *(Sound: Hammering and grinding of pestle and mortar)*
>
> ASPDIN: William! William!
>
> WILLIAM: Yes, father?
>
> ASPDIN: William, go out to the turnpike and get some more limestone powder. They're repairing a road out that way....
>
> MRS. ASPDIN: Joseph, Joseph—how do you expect me to get the dinner if you have our stove filled with dirt and rocks?[20]

Later, the Aspdin character explains to his wife and son how he combines the heated and powdered limestone with water and argillaceous earth (containing clay) to make a mortar that becomes exceptionally hard when dry. There is no explanation of the chemical principles involved, however, and

Aspdin's inventiveness is treated throughout as "puttering" rather than systematic analysis and testing based on sound technical knowledge.

Not every listener (or Smithsonian curator) found the dramatizations enchanting. A California man complained at length about "professional dramatic interpreters" who had, in one episode, endowed the "ancient Aztecs with modern synthetic Hollywood ga ga slush":

> Some person of standing recently declared that modern radio programs are gauged for morons and intellects under twelve years. [The] Smithsonian seems to have accepted this as scientific fact, and the programs seem to [assume] ... that minds above the twelve year grade also like their science diluted with ga ga.... [21]

Other contemporary observers shared this listener's outrage at the Hollywood influence on popular science. And even the Office of Education recognized that scripts sometimes compromised too much in favor of dramatization. "It has come to my attention that frequently our scripts do not contain enough subject matter," one of the directors wrote to Austin Clark:

> You have had experience in radio broadcasting; you know that there must be some compromise between pure science and showmanship. What we would like to do is to put in a script as much "meat" as possible and still keep our scripts interesting to the public.[22]

The desire for positive publicity appears to have stifled such concerns among the Smithsonian leadership. In July 1936 Secretary Abbot wrote to the president of NBC to express thanks for the network's "cooperation and interest," adding that the "great many gratifying responses from radio listeners all over the country ... have convinced us of the potentialities of radio for familiarizing the public with the Smithsonian Institution."[23] Within a few months, over forty thousand listeners had requested supplementary materials from the broadcasts. By March 1939, after 156 broadcasts and three years of production, they were receiving around six thousand letters a week. Secretary Abbot praised the project in a memo to all Smithsonian curators, noting that the institution had experimented with other popularization formats ("lectures, dialogues, questions and answers, etc."), but all these "gradually fell by the wayside, except our dramatized program": "Hundreds of congratulatory letters ... show that the dramatized method is actually the only one that gets and holds nation-wide audiences."[24] Webster True estimated in 1939 that they were reaching "an audience of probably at least 3,000,000," in addition to rebroadcasts within the United States and abroad.[25] Eventually, the Smithsonian received over half a million letters, most of them enthusiastic. When the developers surveyed thirty-five thousand of these correspondents, they found that "the average number of listeners at each radio was four," "the ages ranged in fairly constant proportion from 9 to 60," "slightly more men than women listened," and the audience came from all sorts of occupations and all parts of the United States and Canada.[26] For the Smithsonian leadership, these were encouraging numbers, far greater than the number of visitors to their museums every year.[27] Dramatization was easier to accept if, per Abbot, *The World Is Yours* was assumed to be "the most effective method the Institution has ever used to diffuse knowledge."[28]

Even with elaborate support and enthusiastic responses, *The World Is Yours* was carried by only sixty-two stations in 1937, eighty in 1939, and seventy-five in 1940.[29] As many New Deal projects began to wind down, government funding for *The World Is Yours* was no longer assured. The Smithsonian, unwilling to fund the project single-handedly, asked NBC to pay all production costs if federal funds were unavailable.[30] WPA funding, in fact, was canceled in spring 1940 and publication of the supplementary magazine was discontinued. The radio programs had proved so popular, however, that NBC decided to underwrite part of the production. Starting on July 1, 1940, the Office of Education remained involved in script development while the Smithsonian supported Webster True's time and the salary of a professional writer to produce first-draft scripts. The network paid for final script writing and for the actors, production director, music director, and New York staff.

Like many cultural and arts programs, *The World Is Yours* eventually fell victim to wartime pressures on broadcasting's public affairs and sustaining programs. With many expressions of regret, and over the protests of the Office of Education, NBC canceled the series in 1942. The Smithsonian commitment, it turns out, was already weakening. A year before the cancellation, Secretary Abbot had described *The World Is Yours* in language that hinted at the Smithsonian leadership's growing disenchantment. In a letter to a Canadian scientist seeking advice on a possible radio series, Abbot explained that "educational radio programs, if they are to attract and hold large audiences, must be highly popularized," and only "dramatized" programs were "capable of holding radio audiences of all age groups."[31] The Smithsonian and the Office of Education had made a pragmatic choice to produce a dramatic series because "the listener stays only as long as his interest is sustained—as soon as it lags, he leaves by a twist of the dial." This "ultra-popularization," Abbot continued, had "caused the Smithsonian considerable difficulty in the early days of its radio efforts because the scientists on its staff... were inclined to consider the programs trivial and unworthy of their serious efforts." They had concluded that the best people to produce the dramatized scripts were "expert, highly trained" writers—not Smithsonian scientists. Responsibility for this extreme version of popularization must be handed off to others.

By January 1942 NBC had begun to impose more control over the scripts, sometimes seeking to emphasize aspects other than those approved by the Smithsonian or rejecting proposed topics.[32] Then NBC notified the Smithsonian that the series was being canceled and, despite the institution's overtures to rival network CBS, *The World Is Yours* disappeared from the air altogether in April. The abruptness of the cancellation (and a dispute over the last scripts) spurred a flurry of recriminatory letters between Boutwell and various NBC public service managers. These exchanges call attention to the unresolved issues of control and compromise provoked by "ultrapopularization." Boutwell acknowledged that NBC had a right to participate in content decisions, but not "to cancel the script or touch any part of it in any way": "Only by such policy can the scientific and educational integrity of a script be preserved."[33] We are the experts on education, the Smithsonian curators are the experts on science and history, and no one on the NBC staff (except perhaps the head of all educational programming), he claimed, had "sufficient perspective in either education or science to pass judgment on the values of these scripts." The complaint drew an equally lengthy, inflammatory, and passionate response from two well-educated network managers: "We do not have difficulty... in seeing pretty clearly what are our responsibilities to the audience and in doing our best in spite of tremendous pressure to meet those responsibilities."[34] The scientists and the Smithsonian claimed ownership in the presentation of their ideas and artifacts, but commercial broadcasters controlled the switch.

SCIENCE ORGANIZATIONS, MUSEUMS, AND UNIVERSITIES ON THE MARCH

Throughout the 1930s, other nonprofit institutions, from professional associations to museums to universities, ventured into radio production and tried out some form of dramatic format. None of these series lasted as long (or were as popular) as the DuPont and Smithsonian programs, probably because none enjoyed the same level of financial underwriting. Nevertheless, the breadth of offerings indicates the extent to which academic and nonprofit science sectors came to regard broadcasting as an important form of public outreach.

Thornton Burgess told Austin Clark in August 1935 that since he had acquired a sponsor, he had been "dramatizing incidents whenever possible," although he justified the change as simply "a presentation of Nature in a new way."[35] Clark agreed that the technique seemed to be effective because listeners preferred to have education "conveyed in the form of an amusement," but noted that many of his colleagues objected "to having science portrayed in the press in the form of dramatics or of dollars."[36]

Other scientists, however, were eager to add entertainment elements. When public health expert Homer N. Calver wanted to use sound effects for his February 1935 Science Service radio talk about microbes, CBS suggested a "march effect" (something that would fade out within a minute) and perhaps additional dialogue to enliven the text. As a result, the opening of "The March of the Microbes" declared: "The history of mankind echoes with the endless tread of millions of human beings forever pressing onward to found new civilizations." Then listeners were asked to imagine "the stumbling of hoofs, the patter of paws, the beating of wings," and, finally, the "invisible creatures all around us."[37]

Professional organizations like the American Association for the Advancement of Science (AAAS) were also at the mercy of broadcasters, and began to exploit dramatization in order to secure free airtime. In 1938 AAAS worked with NBC on *Science on the March*. AAAS permanent secretary F. R. Moulton claimed that the series would "advocate confidence in experimentation and the process of reasoning," without presenting science as "magic" or portraying "scientists as heroes."[38] The series featured a Socratic dialogue, encouraging listeners to identify with the program announcer, who then posed questions to a scientist (sometimes played by Moulton or Carroll Lane Fenton, a well-known geologist and science writer). As Moulton explained,

> In Socratic fashion they question and argue, sometimes following false trails, sometimes reaching premature conclusions on insufficient evidence, often triumphing, but sometimes being compelled frankly to admit that they do not know the answers to the questions they are considering. That is the way it is with scientists and with all honest, inquiring minds.[39]

This formula had also been employed successfully by the University of California Radio Service in broadcasts carried regionally via NBC affiliates. The California series featured a character called the "University Explorer," who served as the listeners' surrogate in the interview. The Explorer would describe a science facility and summarize a researcher's work and then introduce and interview the guest.

Moulton's article describing the AAAS project adopted an almost apologetic tone, as if attempting to dispel lingering suspicions of the medium among association members. He also admitted that, almost a year after its inauguration, the series had not achieved its high ideals, the excuse being that radio production was a part-time, amateur activity for AAAS: "It is a somewhat onerous burden to prepare and deliver at a fixed time a broadcast week after week in the midst of many other duties."[40]

Most planetariums, observatories, and museums continued to use conventional formats. At the Franklin Institute in Philadelphia, James Stokley broadcast a "round-by-round description" of a 1935 lunar eclipse on CBS; that year he also appeared again on the Science Service series, providing a lively talk about "Amateur Astronomy."[41] At the American Museum of Natural History's Hayden Planetarium, astronomer Clyde Fisher initiated fifteen-minute *Hayden Planetarium Talks* (carried by CBS, 1936–1939, and then MBS, 1940–1941).

The following year, the American Museum developed a second, more entertainment-oriented series, *New Horizons*, which juggled adventure, exploration, and natural history tales within a dialogue format designed to make schoolchildren feel that they were "overhearing that most exciting of all things, adult conversation on things that are happening in a grown-up world."[42] Museum director (and famed naturalist) Roy Chapman Andrews asked guests to describe their work, and these conversations would be followed by dramatizations. Participants included dinosaur hunter Barnum Brown, naturalist William Beebe, and Arctic explorer Vihjalmur Stefansson. William K. Gregory spoke on "Men versus Monkeys," and Roy W. Miner described the dangers of diving for pearls in the South Seas. Science was infused with romance and beauty, was proffered as adventure rather than laboratory drudgery and rational analysis.

The museum sponsored several other radio series during the 1930s, most for youthful audiences. *Man and the World*, coproduced with Chicago's Museum of Science and Industry, presented a "dramatic panorama of scientific research, exploration, and discovery." *This Wonderful World* centered on a natural history quiz conducted live with schoolchildren who visited the Hayden Planetarium. The planetarium also created a popular fifteen-minute astronomy series, *Men behind the Stars*, coproduced first with Columbia University and later with the Works Progress Administration for the City of New York.

Science was getting caught up in the show. When a science museum was reopened in New York City's Rockefeller Center in February 1936, part of the dedication hoopla involved British scientist Sir William Bragg. Broadcasting from London, Bragg struck a match and lit a candle; the sound was converted to an electrical impulse carried across the Atlantic Ocean via radio and used to light, within the New York museum, an early incandescent lamp, and then to trigger forty new-style mercury vapor lamps. That same NBC broadcast included Albert Einstein in New York and, speaking from California, Robert A. Millikan and Amelia Earhart.[43] Networks began carrying more live broadcasts with scientists—from Copenhagen (physicist Niels Bohr on "International Science") to Chicago (Arthur Holly Compton discussing "Physics and the Future").

Entertainment had been added to the popularizer's bag of acceptable tricks. Following the lead of the Smithsonian and DuPont series, various university broadcasting groups incorporated dialogue, sound effects, and comedy to attract listeners to science. The Field Museum of Natural History, NBC, and University Broadcasting Council (a consortium of Chicago universities) attempted in 1939 to imitate the success of *The World Is Yours* with dramatizations based on museum collections. The promotional brochure for *How Do You Know?* promised "little-known stories about familiar phenomena and others which touch upon the more remote regions of human curiosity."

The dramas in a University of Chicago series, *The Human Adventure*, portrayed universities as not just "storehouses of knowledge" but also investments "in a better life."[44] Carried first by CBS and then by MBS, *The Human Adventure* was coproduced with Chicago station WGN and *Encyclopaedia Britannica* and was created by Sherman H. Dryer, whose duties as university "radio director" included brokering faculty appearances on other radio programs. From 1939 to 1946, the Chicago programs used well-known actors, stunning sound effects and music, and professionally written scripts to explore research "in all fields of knowledge"—from the plague to termites, dental decay, and theories about the origins of the earth. For an episode about Albert Einstein, actor Clifton Fadiman played an "omniscient explorer" and F. Chase Taylor (famous as the popular radio character "Colonel Stoopnagle") starred as a "know-nothing layman." One award-winning episode, called "Typhus," described how DDT was used during World War II, tracing the work of scientists as they identified the typhus disease vector and then developed a chemical weapon to combat it.[45]

Critic Max Wylie praised *The Human Adventure* for tackling complex scientific topics, yet observed that, in the process of simplification, "much has to be sacrificed to be widely understood."

> I doubt if the series, as a whole, has either pleased or satisfied the scientists. But, then, few things please scientists anyhow; they would not be good scientists if it were otherwise.[46]

A program fulfills "its real duty," he continued, if "the information is substantially clear and completely accurate (even if slightly bobbed)," drawing listeners' attention to "work that is important, contemporary, and progressive" and fostering "an active respect for learning and the learned."

Another well-regarded academic series, *Unlimited Horizons*, ran on NBC from 1940 to 1943, with support from the University of California, Stanford University, and the California Institute of Technology. That project also explored the breadth of science—from J. A. Anderson and Edwin P. Hubble discussing "Heavenly Bodies" to Stanford biologists describing conservation of Pacific Coast salmon. Here too the goal was to translate science's "romance" into stories for "the entertainment and edification of Mr. and Mrs. Average Listener."[47]

In series like these, radio popularizers struggled constantly to evoke science's inherent "dramatic meaning" and to describe, as CBS executive Lyman Bryson phrased it, the "impersonality and detachment of the laboratory worker" while still entertaining the audience.[48] It is essential, Bryson wrote, "that the general public gets . . . some sense of the scientists' way of working and . . . way of thinking," even if such an approach requires "compromise" of the "technicalities." The compromises were, in fact, numerous. Dramatic effects ranged from sound effects ("growl and snarl of wild animal . . . orchestra [makes] weird unearthly effect . . . sweeps full to tremolo . . . fades out . . . icy winds sweeping across polar regions"), to an announcer's echoing voice ("For a million years . . . a million years . . . the earth has been frozen in ice. The Ice Age!"), to actors playing "shuffling, shaggy apelike creatures . . . silhouetted against the slate-grey sky."[49] The brief technical statements by scientists within such shows were increasingly like rocks scattered onto snow.

SCIENCE AS "ACTUALITY"

The romantic appeal of drama also influenced news presentations about real scientists and scientific events. Science journalism had long attempted to impart a sense of "actuality" through firsthand accounts of expeditions, such as Science Service's arrangements with explorers and anthropologists. Movie house newsreels, viewed by tens of millions of people every week during the 1930s, increased the cultural confusion whenever dramatic musical scores or narration were used to enliven real documentary footage of scientists or where they worked.[50] From 1939 to 1948, science and health topics constituted between 1 and 2 percent of the content of the top five movie newsreels.[51]

Radio contributed its own special blend of messages to American culture. When *Time* was founded in 1923, the magazine had adapted the movie newsreel's bulletin style to print; then in 1928, *Time*'s general manager, Roy Edward Larsen, created for radio a ten-minute summary of news stories from *Time*'s current issues. *The March of Time* remained on the air until 1945, covering science alongside Hollywood sex and Washington scandal. Larsen pioneered the practice of what he called "newsacting," hiring actors to impersonate real people (reading the statements or speeches of public figures and re-creating their voices).[52] Adding sound effects, music, or authoritative narration was deemed acceptable because all the restagings were based on actual events.[53] Audiences, however, sometimes perceived the re-creations as real, and the practice increased an expectation of drama within all news and current events reporting.

Radio extended the ballyhoo further when listeners were allowed to "join" expeditions in real time. Scientific expeditions offered naturally colorful opportunities for such actuality broadcasts and, beginning in the late 1920s, radio networks scrambled to exploit celebrity explorers. Included in the sixty-man team for explorer Richard Byrd's 1928–1930 Antarctic expedition were *New York Times* correspondent Russell Owen, two motion picture photographers, and three radio operators. Radio served as both an essential lifeline for the expedition and Byrd's platform for publicity. The expedition transmitted over three hundred thousand words of press reports via WHD, a station owned by the *New York Times*.[54] Isaiah Bowman, director of the American Geographical Society, resoundingly praised press attention to the Byrd expedition as good public relations for scientific research:

> The time has come when science as well as the general public should acknowledge its indebtedness to the press. Without the assistance of the newspapers the well-equipped expeditions of recent years could not have been undertaken.[55]

People are "far more interested," Bowman later claimed, "in an expedition from which we can have almost daily radio reports than . . . in one that vanishes for several years, returns with news that blazes for a week, and then drops into a gulf of forgetfulness."

Three years later, using a combination of shortwave hookups, CBS took Americans along on Byrd's next expedition.[56] Sponsored by General Foods, these programs were contrived as "two-way broadcasts," in that listeners heard both scientific reports sent by the expedition via shortwave and the personal messages, music, and entertainment transmitted to the expedition members from home. "The 'back-stage' boundaries of the Byrd Expedition broadcasts are almost limitless," the expedition's newsletter *Little America Times* declared in its October 28, 1934, issue:

> The world is its stage. It is the only program whose producer is 10,000 miles from the "footlights".... The entertainment features of the program are planned well in advance. Two factors must always be considered—what the listeners in the United States want to hear from the expedition and what the expedition wants to hear from the folks at home.

By allowing listeners to "hear Byrd's voice in the comfort of home when he is buried in snow thousands of miles away," Benjamin C. Gruenberg observed, radio could "emphasize the spectacular and exciting."[57] As one person later recalled, "it was a real thrill...to hear someone talk from the South Pole—scratchy as reception was some of the time, a real marvel in those days."[58] Malcolm Davis, Watson's brother and a staff member at the National Zoo, made his second trip to Antarctica in 1940 with another Byrd expedition, and Watson eagerly participated in transmitting radio messages to the expedition via a program sponsored by newspapers around the country.[59]

Even when actuality broadcasts involved ongoing research projects, the science had to adjust to radio's time constraints. In 1932 NBC asked William Beebe, director of tropical research at the New York Zoological Society, if they could broadcast from his bathysphere when the biologist made his next descent off the coast of Bermuda. Beebe was already a public figure; his earlier dives had been featured in movie newsreels, and he was by then a prolific popularizer via books, magazine articles, and lectures. After several weeks of preparation by the broadcasting team, during a spate of bad weather that included two hurricanes, the group attempted the dive on several Sundays in succession. Each time, the radio audience would be alerted. Each time, equipment failure or dangerous swells caused cancellation. Finally, on September 22, "with a sea still too rough for comfort," Beebe agreed to risk a dive rather than default on the arrangements:

> With only two hours' notice, messages went out to New York and were relayed to scores of American and foreign stations. Singers and entertainers of all kinds were switched off, and the wires cleared for this new experiment.[60]

The broadcast was divided into two half-hour segments. During the first part, the NBC announcer described the activity on deck while Beebe was sealed into the bathysphere. Sound effects heightened the drama:

> Two sound microphones, mounted near the bathysphere and the big winch, caught the clanging of the sledge hammers tightening the nuts of the door, and the grinding of the winch as it released more and more cable to lower the bathysphere into the deep.[61]

The announcer encouraged listeners to regard themselves as "ex officio members of the expedition" and outlined the dangers Beebe would face. The announcer also attempted to refute criticisms of such stunts:

> This is a broadcast of a scientific undertaking. It is *not* a planned event to make a radio program. There are too many attendant dangers and the exploration is too important, to forecast now just exactly what or how anything may happen.[62]

Audiences from around the world tuned in while the naturalist dropped twenty-two hundred feet and described the undersea life visible through the vessel's thick window (signals were relayed to the United States and by shortwave to England for rebroadcasting on the BBC).[63] "Beebe was an exceptionally fine broadcaster," one contemporary wrote, because he made the listeners feel "in a most tangible fashion the confining walls of the diving ball as it was lowered from the surface water a half mile."[64] Because of the BBC rebroadcast, Beebe biographer Robert Welker points out, the scientist had achieved an audience "greater than any single lecture . . . would reach, and under circumstances more dramatic than any of these could command."[65]

Use of radio to carry listeners to research sites continued to be popular throughout the 1930s. In 1937 astronomer S. A. Mitchell led an expedition to the South Pacific Ocean, sponsored by the National Geographic Society and the U.S. Navy, to observe a total solar eclipse. Along with scientists, photographers, and artist Charles Bittinger (a longtime friend of Watson Davis), the group included two NBC radio engineers who arranged over a dozen network broadcasts documenting the preparations, progress, and eclipse observations.[66] CBS sent two radio personnel on the Hayden Planetarium's expedition to the Peruvian Andes to observe the same solar eclipse and to broadcast interviews from the site.[67] The next year, NBC arranged the first broadcasts from the site of archeological digs at the pyramids in Egypt, interviewing George Andrew Reisner and his colleagues inside the tomb of Cheops.[68] On other occasions, naturalists carried NBC radio equipment into the Amazon jungle so they could thrill audiences with highlights of the Terry-Holden expedition, sponsored by the American Museum of Natural History. Broadcasts were also made from the MacGregor expedition to the Arctic Circle.[69]

UNLEASHING THE IMAGINATION

These actuality broadcasts involving scientific expeditions surged into listeners' living rooms amid many other voices—including those of fictional detectives who adopted "scientific techniques" to solve crimes, space heroes who used science to save the universe, and fictional doctors who promised that science would cure a soap opera heroine's terrible disease. Throughout popular culture, science was being incorporated cheerfully, creatively, and covertly. Science was treated with a familiarity that was simultaneously hopeful, respectful, and skeptical. Hollywood film biographies of Marie Curie and Louis Pasteur provided inspiration; newsreels dramatized the latest discoveries in astronomy or physics; and within a colorful array of popular novels, comic books, and movies, science constantly came to the rescue. From *Jack Armstrong, All-American Boy* (1933-1951) to *Buck Rogers in the 25th Century* (1932-1947) and *Flash Gordon* (1935-1936), radio's heroes exploited scientific tricks to defeat their opponents. These positive messages about science were reinforced each week at the neighborhood movie theater. The two most popular movie features of 1938 were *Flash Gordon's Trip to Mars* and the serial *Mars Attacks the Earth!*[70]

Incorporation of science (both real and imaginary) in one genre sparked imitation in others. Comedian Jack Benny offered his version of scientists' search for a yellow fever cure in a freewheeling radio adaptation of the Hollywood film *Yellow Jack*, which was itself based on a successful Broadway play cowritten by microbiologist Paul de Kruif.[71] The success of movie versions of Mary Shelley's novel *Frankenstein* spawned cartoons and comic books about "mad scientists" and radio dramas about reckless experimentation.[72]

Some observers warned that even the best-written dramas, and even those with positive messages based on real events and discoveries, could potentially confuse the lay audience. In 1936 Watson Davis told chemist H. E. Howe that, despite notable scientific errors, he believed that the movie *The Story of Louis Pasteur* would "do a great deal of good in getting across to the public the spirit of science." Audiences should, however, be better informed about dramatization:

One way of injecting into a motion picture the dramatic interest the producers seem to require, when that dramatic interest is not in accordance with historical fact, and yet keep faith with truth, would be for the producer to state... at the beginning that certain fictional sequences have been introduced... for dramatic effect but that... the spirit and essential facts are correct.[73]

No radio presentation used popular culture's amalgamation of fictional and real science to such advantage as the 1938 dramatization of H. G. Wells's *War of the Worlds*.[74] The presentation and the reaction to the program encapsulated all the pitfalls and challenges of this evolving "ultra-popularization" era. The novel chronicles an invasion of England by Martians whose spacecraft are equipped with terrible "death rays." It is a clever, compelling book, rich with social commentary. For the October 30 installment of his weekly CBS drama series, *Mercury Theatre on the Air*, the actor and director Orson Welles reset the plot in 1930s America, landed the Martians in New Jersey, abandoned the narrator character, and revealed events through a sequence of simulated news bulletins, "eyewitness" descriptions, "on-location accounts" by reporters, and "interviews" with experts and local authorities, all delivered by actors, who were ostensibly preempting an orchestra concert. Whenever the dance music was interrupted for a report from the scene of the "invasion," sound effects such as crowd noise and police sirens increased the dramatic effect. For the fake news bulletins, the actors copied the style and language of contemporary radio announcers.

By mimicking 1930s actuality broadcasts, Welles intended to heighten the impression of realism. He did it so well that some listeners believed that they were hearing real news reports. The next day, one newspaper wrote, "A lot of Washingtonians and other Americans probably felt sheepish... over being badly frightened... by a too realistic radio program."[75]

Welles's script exploited the authority with which radio conventionally treated science and scientists. A CBS survey found that 42 percent of those who believed they were hearing an actual news broadcast had tuned in late and missed the opening disclaimer.[76] To them, the faux interviews with scientists sounded convincing. The actors peppered their comments with technical jargon not all that different from what listeners were accustomed to hearing whenever Watson Davis interviewed scientists on the Science Service series. Interviews with "astronomers" from the "Princeton Observatory," who reported seeing explosions on the surface of Mars, were followed by reports that scientists at the "California Astronomical Society" disagreed with the first group's observations.[77] Neither institution was real, but again the names were plausible.

By coincidence, the broadcast took place just as a group of social scientists was launching one of the first comprehensive studies of radio's social impacts. Hadley Cantril quickly obtained additional funding from the Federal Radio Education Committee and the Rockefeller Foundation to study the Welles program and to interview a sample of listeners. The audience for *War of the Worlds* was large—some have estimated as many as six million people eventually tuned in. At least one million of these, Cantril concluded, became "frightened or disturbed" by the broadcast.[78] Some newspapers reported that there was considerable panic in parts of New Jersey near the purported landing spot. Cantril's research found that, while over 20 percent of those who claimed to have been frightened had actually listened to the broadcast from the beginning, most of that group lived, perhaps predictably, near where the Martians had supposedly landed. There was a measurable increase in the volume of telephone calls to police stations in that area during the broadcast, and a 39 percent increase in telephone use overall in that area of New Jersey.[79] The network's switchboard was so flooded with calls that Welles was compelled to break into the drama to remind listeners that this was merely a play.

The extent of public agitation over the broadcast can often seem puzzling to contemporary readers. Some historians have pointed out that Americans were becoming accustomed in the late 1930s to receiving news (especially breaking news relating to the European conflict) via radio. Welles had mimicked this format with extraordinary skill. The bulletins, the threat of invasion, and the descriptions of violence also tapped into simmering prewar anxiety among East Coast residents.[80]

Popular science had also prepared the human imagination for at least the possibility of life on Mars. Discussion of Mars was not unusual on radio, in fictional and factual presentations, in the heroic adventures of Buck Rogers and Flash Gordon and in the serious Science Service talks. As Cantril observed,

> For many persons another bewildering characteristic of our present civilization is the mystery of science . . . a world outside and . . . a universe of discourse completely foreign to the perplexed layman. . . . If science can create the things we have, why can't it create rocket ships and death rays?[81]

The incorporation of science within and by popular culture offered useful opportunities to grab the interest of ordinary people, to demonstrate science's excitement, creativity, and, yes, occasional flashes of drama. On radio, where the pressure of time compressed all text and repelled all efforts to qualify and annotate, that incorporation posed a danger. The context could also easily obscure error and irresponsible fictionalization. As Louis Reid observed at the time, "Symphony and swing, politics and hog-feeding, comedy and propaganda, dictators and democrats, news and Tin Pan Alley, Shakespeare and Broadway, science and ballyhoo—jumble them together and you have radio."[82]

During the 1930s, the creators of radio drama experimented with "dramatic speech," encouraging listeners' willingness to suspend disbelief and engage in fanciful alterations of space, time, and history.[83] Where, then, did reality lie? What did it matter if radio dramas mixed the actual words, names, theories, discoveries, and insights of scientists with hypothetical or fanciful versions of the same? And why should serious science programs on radio not add hints of drama, or even little dramatizations? The difficulty lay in the absence of disclaimers, in the extent and intent of deception, and in how well audiences had been prepared to evaluate what they heard. How could ordinary listeners detect the fiction? Who was helping them learn how to distinguish between fact and fiction? Here was an issue that went beyond scientists' traditional concern with the accuracy of popular texts, beyond preoccupation with whether terms were pronounced correctly or numbers rounded off. As more and more scientific organizations jumped on the drama bandwagon, responsible popularizers like Watson Davis wondered how they could keep audiences interested yet still "keep faith with truth."

CHAPTER EIGHT

———— ✳ ————

Adventuring with Scientists

Science consists of (1) impersonal and (2) timeless truths.

E. E. SLOSSON, 1927[1]

FROM the outset, Science Service's radio presentations had centered on the science rather than the scientist. The presenter, whether a guest scientist who read a talk or a journalist who summarized the research and introduced the guest, had been chosen to lend authority, not human interest. Programs described the essence of research, outlined the state of knowledge in a field, shared insights, and sometimes related science to economic development, but gave only limited attention to the human component. Even when opening news segments were added, the rhetorical focus remained on the corporate achievements of science, not the individual triumphs.

In 1934, reflecting trends within radio generally, Watson Davis began to introduce more personality to the Science Service programs. The next year, "at the request of the network," he added interviews. Three years later, CBS renamed the series *Adventures in Science* and initiated a format that remained unchanged for the next two decades: an opening sequence of science news bulletins and a scripted conversation with a guest.

From the mid-1930s on, scientists rarely appeared on the Science Service or similar radio shows without an intermediary who introduced and summarized the topic, asked questions, and generally kept a program moving. The arrangers diplomatically negotiated the conflicting interests of scientists and networks, placated scientists' egos, and worked diligently to overcome distrust of a "sensationalistic" medium. Scripts were also shaped to meet the goals of the scientists' employers (which included ever more industrial laboratories) or public relations representatives, while titles, topics, and text were adjusted to prevent listeners' tuning out.

By then, scientists were also being acknowledged as sources of "news" rather than merely the producers of fascinating, timeless information. Announcements of spectacular discoveries in fields like physics and archeology added to a sense of excitement. Yet when CBS executives began to pressure Davis to add "spot news," he initially hesitated, lest the change damage Science Service's relationship with the scientific community.

Journalists and scientists engaged in a carefully structured dance. Davis, who depended upon researchers as news sources, had won acceptance (and gained cooperation) by exhibiting respect for the scientists' norms of accuracy, appropriateness, and credit. He and other major science reporters had attempted to honor researchers' requests that results never be announced in the press before journal publication or, at least, before formal presentation to scientific peers. During the mid-1930s, however, competition among media outlets intensified. Radio broadcasters felt compelled to compete with newspapers for "scoops," and newspaper publishers felt squeezed by radio's ability to broadcast bulletins as soon as they sped over the wire services. The long-standing informal rules for *all* news reporting began to be reassessed, with consequent effect on the treatment of science.

The evolution of the "Science Service Talks" into *Adventures in Science* reflects the trends affecting all science popularization at the time: more emphasis on exciting breakthroughs rather than patient accumulation of knowledge; increased attention to social issues; a shift of focus from the science to the scientists; and the elevation of prominent scientists to the status of cultural celebrities. The remodeling of *Adventures in Science* also demonstrates how popular science became entangled with the commercial business of delivering mass culture. In 1938 and 1939 Watson Davis experienced a brutal reminder of who controlled radio. All the good intentions in the world—including noble visions of bringing science to the public—were so much stardust without access to a microphone.

GETTING THE NEWS ON THE AIR

But first—the news. Science journalists like Davis asked and answered the question "What's new?" far differently than a researcher in the laboratory. Yet, because they effectively served as intermediaries between that researcher and the general public, they also approached the question differently than might other media professionals. Around 1934 Davis began discussions with NBC about creation of a news-oriented science program. Those negotiations accentuated the differences in how scientists and the media perceived the issue of "timeliness."

To a media professional, the value of news is in inverse proportion to its age. The typical newspaper editor, E. E. Slosson emphasized to his trustees, is "a Mendelist, not a Darwinian": "He believes in evolution by jerks not in gradual development by imperceptible degrees."[2] That editor does not feel the need to wait for the confirmation, interpretation, and validation of scientific journal publication; a scientist's professional assurance of accuracy will often be regarded as sufficient. Scientific progress, however, results from "the slow accretions of long continued labor," at the end of which is publication. Ordinary scientists see providing information to a newspaper before results are published in an accredited scientific journal as rushing into print, as pandering to sordid sensationalism, and as alien to the traditional practices of responsible researchers.

To the radio network executives, the recent triumphs and spectacular discoveries of scientists made a news-oriented science show attractive, and Science Service was in a favorable position to develop such a program. Its extensive network of stringers and supporters could provide tips and could help explain the latest, most complex research.

As physicists raced to understand the basic structure of the atom, for example, journalists sometimes stood like spectators at a Mardi Gras parade, trying to make sense of the floats and catch bits of news tossed by the participants. It was helpful to have friends among the cognoscenti. "How did you pass up the experiment of [Enrico] Fermi which is described in the sheets which I enclose?" California Institute of Technology professor R. M. Langer wrote to Davis in April 1934: "They are a translation of his first notes which just arrived here (private communication). I was tempted to wire you, but I thought that by this time you must have the news."[3] Langer, a regular contributor to the news service, then enclosed his succinct, articulate summary of the implications of Fermi's article "Radio-Activity Induced by the Bombardment of Neutrons," recently published in an Italian scientific journal.

In competitive, publicity-conscious fields like paleontology and archeology, Science Service had even been able to capture some scientific news as it was being made. One of the early fossil finds of Charles Lewis Gazin, a scientist at the U.S. National Museum, had been the subject of a Science Service *Mail Report* in 1933. In July 1934, when a Smithsonian colleagues of Gazin's gossiped about another spectacular discovery, Davis wrote to the vertebrate paleontologist at his Idaho dig.[4] Davis enclosed an instruction sheet, "Writing for Science Service," and prepaid Western Union blanks. Gazin at first replied that he had nothing to report, but then sent enough information for a *Daily Mail Report* story on August 16, 1934, describing the uncovering of Pliocene period remains of the genus *Plesippus* ("U.S. National Museum Gets 2,000,000-Year-Old Horses").[5] The cordial relationship with Gazin lasted for years; there were thirteen more *Daily Mail Report* stories on his work before the scientist enlisted in 1941, and coverage resumed after the war. Such experiences exemplified how Science Service journalism differed from that of the normal beat reporter. Davis and his staff were constantly balancing the goal of getting news with the need to maintain cozy relations, even friendships, with scientists who might be sources in the future. At the heart of this relationship was constant sensitivity to timing, accuracy, and credit.

What was "news" in science was not always new. As Davis attempted to explain to an NBC staff member, "Most of the science news does not break in the way that a murder or shipwreck or other news of that character happens. It is a little more deliberate."[6] Journalists might prefer not to wait for scientists to release results on their own timetable (which could give the appearance of staleness), but the researchers' cooperation in assisting the news process was vital. Science Service staff worked hard to convince their sources that, while they valued accuracy over haste, there were also deadlines to be met. Cooperation in advance would insure a more accurate story later. Davis acknowledged the challenges:

> News of science does not develop like news of war, politics, crime and sport. Practically all scientific news is the result of months or even years of patient research, and it is produced by men who would rather remain silent than make an announcement that was not thoroughly authentic.[7]

Who would obtain the latest news about science turned out to be a key factor in the NBC negotiations. Margaret Cuthbert, a network education specialist, wanted to originate a new sustaining program (to be called *Do You Know?*) from Washington, D.C., based on "the most universally interesting" stories from the *Science News Letter*. "The whole point in doing this," Cuthbert wrote to Davis, "would be to have the broadcast the day before the News Letter comes out so that the papers would comment upon your broadcast editorially as news."[8]

Davis toyed with the idea, first suggesting a format of "about 8 minutes of vivid, colorful descriptions of just what a scientist's work means" to be followed by an interview with that scientist. Davis claimed that he could easily obtain the participation of "eminent scientists" like "Millikan, Compton, Abbot," and he indicated a willingness to explore the "more ambitious and time-taking approach" of "semi-dramatization of scientific accomplishments."[9] Cuthbert rejected that idea but continued to press Davis on some sort of science bulletin service, suggesting that perhaps he or another member of the staff would, on short notice, do "a spot broadcast that has news value" and then interview the scientist involved.[10] Davis countered by placing the initiative back on the network: "When a science event comes into the news that you think would make such a broadcast you wire me and we shall see what can be done."[11] Cuthbert wanted Davis to do the work because he knew the world of science best: "with your knowledge of forthcoming events you can get it on the air before it breaks in the papers."[12] Davis declined diplomatically. He knew that such an activity (for which Science Service would receive no revenue) would, in effect, scoop the organization's own newspaper clients and damage its reputation for fairness. In December 1935 Davis flirted again with the possibility of setting up a "tip service" for NBC but eventually rejected the network's proposal.

WHOSE STANDARDS?

NBC's offers testified to growing public interest in science news, something further evidenced by the founding in 1934 of the National Association of Science Writers (NASW). The number of specialized journalists had grown as newspapers and other press associations gave more attention to science. During the worst of economic hard times, and despite the failures and consolidations of other news syndicates, Science Service still retained most of its clients because, Davis wrote later, "a good many newspapers did not consider science a luxury that could be dispensed with under the influence of financial pinching conditions of the depression."[13]

Disagreements that arose during the establishment of NASW demonstrated, however, that these writers' views were not homogeneous. "Science journalism" encompassed a range of perspectives, approaches, and allegiances. In June 1934 David Dietz, Howard W. Blakeslee, William L. Laurence, Gobind Behari Lal, and John J. O'Neill asked medical reporter Jane Stafford to join their discussions on behalf of the Science Service staff. Dietz was science editor for the Scripps-Howard News Service; Blakeslee, science editor for the Associated Press; Laurence, science news editor of the *New York Times*; Lal, science editor for Hearst Newspapers; and O'Neill, science editor of the *New York Herald-Tribune*. The five men had just elected

Dietz "president" of what they called a "press association," and they wanted Stafford to persuade Davis to participate. Stafford told Davis that she thought that NASW was "a lot of foolishness" and "an outlet for David Dietz's ego."[14] Her skepticism was echoed by Davis and Thone, who considered the proposed NASW standards for ethics and performance to be looser than those already embraced by Science Service. As Davis explained, "I believe Science Service ethics methods are and must be higher than any code adoptable by [the] science writers organization and we should not lend our support and approval to individuals and practices over which we have no control."[15] After the group's objectives were adjusted to eliminate what Davis called "political logrolling" and the "trade union angle," he endorsed the charter and agreed to become involved, but Thone refused to join for almost a decade.

The concerns expressed by Davis and Thone reflected the closeness of their ties to the scientific establishment, which were far more convoluted than those of ordinary newspaper reporters. Science Service continually played a "mediating" role. Its staff worked directly with individual scientists to develop and draft articles, and then to syndicate that material to newspapers. Since the organization's founding, it had also strived to develop a model for popularization and to help all scientific information flow easily from experts to public. As a result, the Science Service journalists tended to perceive and assess their work in a context beyond their own organization's self-interest. Increased errors would damage their relations with sources but could also erode public confidence in science; unwarranted emphasis on "break-throughs" might discourage the cooperation of responsible scientists and also encourage sensationalism. Science Service reporters therefore had far more conservative attitudes toward the timing of the release of research results than their competitors.

Other science journalists did not always exercise such concern or caution. The newspapers' appetites for information were insatiable, competition was growing, and attitudes toward news gathering were increasingly aggressive. In the mid-1930s, some major science writers began to pressure the professional associations to increase the "spoon feeding" of journalists. William L. Laurence suggested that physicists who submitted technical abstracts for major meetings might be encouraged to write additional popular versions for the press in an effort to reduce reporting errors; David Dietz called on the new AAAS president, Karl T. Compton, to require that an abstract or copy of every paper delivered at association meetings be provided to the press bureau; and Howard W. Blakeslee urged the American Psychological Association to do the same, citing AAAS as the model.[16]

From the scientists' perspective, the question was the opposite: not whether they should relinquish more power to the journalists but how they could achieve more control over popular content. Various policies were proposed or already in place for managing when and to whom results would be released at major scientific meetings.[17] When AAAS considered establishment of an informal press review board, Davis complained to Austin Clark that the suggestion had "a fascist aroma."

> It boils down to the question of whether there is to be freedom in science reporting, freedom which is made possible by the competence of science reporting, or whether there is to be such incompetency in science reporting that Boards of Review or censors, if you wish to call them that, will be necessary for the protection of scientific men.[18]

Many university laboratories already had policies that controlled the timing of scientific announcements or use of material from research. Wellesley College astronomer John C. Duncan, for example, informed Robert Potter that he could not supply the newest photo of a ring nebula because Mount Wilson Observatory, where he had conducted the research, would not "issue photographs for general publication until after they have appeared in a technical journal accompanied by the author's name."[19] Not one to give up on a good story, Potter pressured the observatory director directly, explaining that they needed an earlier release to get the photo into the production process and promising that they would not "break the release date."[20]

ADDING "PERSONALITY"

Davis flirted with NBC throughout the 1930s, but despite the fact that he was a childhood friend of the network's president, Lenox R. Lohr, the journalist never consummated a deal for the proposed news series. The request involved more than Davis wanted to deliver. NBC executives desired both scientific "news" and "the personality himself or herself," Margaret Cuthbert explained, with Davis "bringing the person to the microphone, introducing them and framing the picture for them."[21]

Although such a project never moved forward on NBC, Davis did eventually revamp the Science Service series on CBS to increase its human component, first by simply using remote broadcasts to tap "the reservoir of science" beyond the Washington area and later by doing what NBC had originally sought.[22] Shifting attention toward the scientist had long been suggested by various trustees and advisors ("People are not interested in things alone. People are interested in people," one wrote).[23] Slosson had always resisted that approach, explaining that "the science we are trying to get over has just the opposite aims" of the person-centered approach, for "science consists of (1) impersonal and (2) timeless truths."[24] Now Davis moved the radio series away from Slosson's idealized model and toward the personal and the timely.

In fact, Science Service content had never really ignored the scientific "stars." The sixth issue of *Science News Bulletin* (May 9, 1921) included stories on Albert Einstein's lectures at Princeton and the anticipated U.S. visit of Marie Curie. By the 1930s, however, an emerging cult of the celebrity, oriented toward the elevation of public personalities and fascination with entertainers, coincided with scientists' growing cooperation in popularization and their awareness of its public nature. When they consented to be interviewed on the air, researchers now sometimes responded by saying how happy they were to join "the other men of science" on the radio.

For the radio programs, Science Service staff used stringers to identify people who might be invited and also worked with university and industry public relations staff to negotiate scripts. The process required diplomacy, patience, and ingenuity. Davis, Thone, and the others learned how to push for changes yet compromise gracefully to keep a scientist from balking at the last minute. The University of California News Service, for example, had arranged talks in spring 1935 by various faculty members, including E. R. Hedrick, a professor of mathematics. Hedrick's first draft adopted a "pugnacious" and "belligerent" tone and argued that the public (and school systems) had inadequate respect for mathematicians and mathematics (the "Queen of the Sciences"). Thone (who knew the rarity of popular talks on mathematics) worked patiently to persuade Hedrick to tone down the rhetoric and produce an acceptable script.[25]

Witty titles—which CBS regarded as crucial for attracting listeners—sometimes provoked stiff objections from guests. Davis attempted to persuade one archeologist to change a proposed title:

> A short, catchy title to attract the interest of the general public would be more appropriate than "Why Archeology." Although short, this title does not give the general listener any indication as to what the discussion might cover and titles for radio broadcasts serve primarily as "bait" to catch the interest of the listener.[26]

Davis had suggested "Digging Buried Treasure" or "Digging Up History," but the scientist insisted on "The Lure of Archeology." Another characteristic exchange preceded an appearance that DuPont brokered for one of its chemists. Henry J. Wing proposed the scintillating title "Application of Research in the Protective Coating Industry" for his talk about varnishes and lacquers. Davis suggested "Vanishing Varnishes," and included in his letter the same sentence about "bait" for listeners that he had used earlier in writing to the archeologist. Wing refused. While admitting that the suggestion was "certainly 'snappy' " and "attention-arresting," he offered a compromise— "Changing Varnishes"—as "just as suitable and perhaps more accurate."[27]

Even scientists whom they had interviewed for years in print could balk at a radio appearance. Plant physiologist Philip R. White, invited to discuss his latest research, had gotten the approval of his supervisors at the Rockefeller Institute for Medical Science at Princeton, but remained uneasy. White was concerned that his results might be overstated ("Wild publicity is a very dangerous thing") and suggested that he might simply discuss older research in the broadcast instead of the new work on root tissue culture that had attracted newspaper coverage.[28]

ADDING DIALOGUE AND ADJUSTING TO SEASONAL EFFECTS

The "intrusion" of the human element could not be forestalled. Popularizers were being forced to accommodate to the demands of those who owned the platforms, although sometimes one of the experts took the lead. In July 1935 Fred O. Tonney, director of research at the Chicago Board of Health (who was also the American Public Health Association's "director of radio publicity"), persuaded Davis to let him use a dialogue script, in which Tonney took a young woman (described on the show as "a new recruit" to his technical staff) on an imaginary tour of a laboratory. The broadcast—"The Public Health Laboratory—What It Means to Mr. and Mrs. Citizen"—originated from a Chicago studio and employed a number of sound effects.[29]

That fall CBS specifically requested that Davis enhance the "attractiveness and effectiveness" of the series by introducing a "dialog between the director and the men of science."[30] Programs would begin with an introduction and lengthy description of the subjects to be discussed; that is, before the audience heard the scientist, they would hear an intermediary's interpretation of the scientist's work. "We are attempting to make these programs something more than simply a talk, and we have therefore begun to write them in the form of a dialogue," Davis explained.[31] Science Service staff writer Emily C. Davis assumed a greater role in writing script continuity, interviewing potential guests by phone and in person, obtaining background material, occasionally reworking complete scripts, and suggesting topics that might be emphasized.[32] Watson Davis gave her feedback, edited the final draft, and then conducted the interview on air.

The organization's liaison at CBS during this transitional period was twenty-seven-year-old Edward R. Murrow, the newly appointed director of radio talks, and his assistant Helen J. Sioussat.[33] Murrow expressed "enthusiasm" for the proposed changes, calling them a marked improvement over the old format. He also (understandably) suggested eliminating such sound effects as "war whoops and drums" at the beginning of the December 3, 1935, script titled "America 8000 B.C."

> ANNOUNCER—This is the 285th Science Service program which each Tuesday afternoon takes you to the Land of Science. Mr. Davis, where do we go today?
>
> WD [Watson Davis]—About 10,000 years in the past, to the days when America was young.
>
> ANNOUNCER—Ah, indians! [Murrow eliminated the sound effects here.]
>
> WD—Yes, indians, but older than the kind of Indians you are thinking of, in war paint and feathers. We are journeying back to meet the first settlers of America— and that means the first discoverers of the land we live in.
>
> ANNOUNCER—The first Americans, eh?...Twang your bow, Mr. Davis, the hunting ground is yours.[34]

The broadcast continued with Davis's chatty discussion of the latest research ("Until recently cautious men of science would have raised their eyebrows and smiled, if you had argued that America was inhabited earlier than, say, three or four thousand years ago"), and then archeologist Edgar B. Howard described his discoveries at the Clovis site in New Mexico.

The series began to adopt a new tone, sometimes lighter and ever more sensitive to science's social and political context. In a letter chiding one of Murrow's staff for using the wrong title in publicity, Davis added at the end that a forthcoming program ("Running Horses," with the Carnegie Institution's H. H. Laughlin) would be "quite exciting and may even cause those of us who bet on the races to listen in once for a science program."[35] "Crime Laboratories," a program about bullet comparison and handwriting research at the U.S. Bureau of Standards, referenced the detective dramas becoming popular throughout radio (and foreshadowed today's entertainment attention to forensic science). Other programs demonstrated increased attention to conservation, such as the talk by W. B. Bell, chief of the Division of Wildlife Research, U.S. Biological Survey, about "When the Ducks Fly South."

That December, when zoologist Henry B. Ward talked about "Pure Water," the program demonstrated how science popularization could be used to encourage political and social change. Ward, a professor at the University of Illinois, was an expert on fisheries and parasites and a past president of the Isaak Walton League of America. He held passionate opinions about water problems in the United States. Pollution was "a menace to the welfare and even the existence of the American people," and Ward wanted to make strong political statements to that effect in the script. He especially deplored the industrial and residential practices that were pouring untreated factory waste and sewer effluent directly into the nation's rivers and streams: "The waste from Chicago's sewers and factories have transformed 200 miles of the Illinois River—one of the most beautiful rivers in America—into a stream entirely without life except the germs of pollution and decay." The effects on waterfowl and migratory birds, fish, and humans were preventable. As he described example after example of polluted waterways—from the Hudson River near his boyhood home in upstate New York, to Washington's Rock Creek Park—Ward spread the blame widely. Governments were failing to enforce existing laws and implement tougher regulation; taxpayers resisted investing in treatment plants; and "Every little laundry pours chemicals into the stream of waste from the city. Every little garage pours in oil waste."[36]

The interaction between scientist and interviewer established a compelling environmental litany, each technical fact matched by an emotional reference to "the beautiful, valuable waters of America" whose purity and serenity were threatened by foul pollution. The dialogue format added to the dramatic effect, introducing a sense of the scientist's passion for his work and a note of urgency to the political message.

Both network executives and audiences praised the format change. From then on, the Science Service programs included carefully scripted discussions, and guests rarely spoke for long without interruption.

Programs strived to link science to listeners' daily concerns. In February guests discussed conservation ("A Future for America's Birds and Beasts"), epidemiology ("The Geography of Disease"), psychology ("The Criminal Mind"), and the ever-popular topic of weather prediction. They also attempted to push the boundaries of acceptable radio subjects. New York state public health commissioner Thomas Parran Jr. was invited to talk about syphilis prevention and treatment (a topic Science Service had already discussed in its print reports) but Davis could not persuade CBS to overcome its taboo against mentioning the word on the air.[37] Less controversial but relevant topics were developed with the cooperation of government agencies like the U.S. Department of Agriculture (USDA). R. G. Webb's July 1936 discussion of an improved strain of cotton exemplified how Science Service used and reused interesting material from other media formats. That program combined text from a script disseminated by the USDA Radio Service ("Strong Hopi Cotton"), a radio talk delivered on USDA's *National Farm and Home Hour* ("Recent Cotton Research and Results"), and a Science Service news feature ("Hopi Cotton").[38]

Davis relished the opportunity to humanize scientists with "dialogue, dramatic programs with music, as well as other types of programs written for the ear instead of the eye."[39] He also attempted a little humor. When he interviewed Charles C. Concannon, chief of the Chemical Division at the U.S. Department of Commerce, about the generally uninspiring topic of tung oil manufacturing and its use in waterproofing, Davis opened with the chipper observation "There are a lot of C's in that name of yours, Mr. Concannon." To which the chemist replied, per the script, "Yes, and in chemicals and commerce. But there aren't any C's at all in tung oil, and perhaps I'd better start by spelling it."[40] This lame joke was followed by similar banter, as Davis asked why the substance was important, where tung oil came from, and what it was used for. They gamely attempted to inject some spirit into a substance primarily important because it helped paint dry.

Even with bursts of humor or occasional sound effects, science still could not compete with baseball in the spring or football in the fall. As Davis explained, "Unfortunately, science . . . sometimes must give way to events of less intrinsic value."[41] Summer political conventions, battleship dedications, Memorial Day automobile races, politicians' speeches, the Poughkeepsie Regatta, and the vagaries of rain delays all provoked elaborate contingency plans. Thone complained about even missing a broadcast interview with an old friend, biologist Anselm Keefe:

> We didn't get to hear you. . . . I wanted to, and the bunch down the hall . . . were even more anxious to tune in. [But] the local CBS station had sold the time to a commercial sponsor to broadcast the baseball game, so you were crowded out. I never before saw the baseball fans in this office praying for rain at game time.[42]

Radio, like newspapers, had also begun to experience what editor Thomas R. Henry termed the "weight of tradition," pressured to include seasonal topics.[43] Thone delayed an invitation to an expert on pollen research until fall because "popular interest in hay fever rises to a peak when the ragweed pollen begins."[44] Emily Davis asked a U.S. Weather Bureau scientist to use the title "Speaking of the Weather" because "people do talk about the weather so much in hot seasons."[45] Programs on "That Perennial Public Enemy,

Poison Ivy" consistently attracted listeners. And food science proved another success. "Bigger and Better Blueberries," featuring USDA botanists and home economists, exploited that juicy topic in June 1936, just as the berries were coming on the market.

SMASHED ATOMS AND SENSATIONALIZED SCIENCE

By the end of 1937, CBS had become the largest single radio network in the world, with 114 station affiliates. As with the other networks, the CBS schedule emphasized both highly profitable sponsored programs and sustaining series like the Science Service programs. The latter were always presented at a network's pleasure. Sustaining broadcasts involved not only a network's "broadcasting time and facilities, but, to a large extent, considerable expenditures of time and effort in the preparation, production and broadcasting of the presentation."[46] Nevertheless, well-produced, interesting sustaining programs could deliver audiences to the sponsored programs, so CBS began to impose more quality control on its educational and sustaining programs, insisting that educational presentations "ought to entertain as well as instruct."[47]

The dialogue format worked so well for the Science Service series that, rather than return to the easier-to-manage radio talk format while he was in Europe during summer 1937, Davis arranged for substitute interviewers. The following spring, he began discussions with Sioussat and her new boss, Sterling Fisher, about how the series might be further improved, perhaps by adding sound effects at the beginning and end or changing the title to convey drama and excitement.[48] Throughout these discussions, Fisher held out a carrot—the possibility of an evening time slot. Radio was then reaching into twenty-six million American homes (and about five million automobiles); network surveys indicated that the potential audience exceeded seventy-five million people.[49]

The series was renamed *Adventures in Science*, in tune with rhetorical flourishes found elsewhere in popular culture. Fisher took a hand in shaping topics and guests, as indicated in a telegram to Davis:

GREATLY NEED SAMPLE SCRIPT OUR PROJECTED SERIES NOT LATER THAN MONDAY AM. . . . SEE STRONG POSSIBILITY WE CAN PUT THIS SERIES IN OUR ADULT EDUCATION EVENING PERIOD. ASSUME FIRST ON SMASHING ATOM SO ASKING ROBERT LYND PREPARE THREE MINUTE SOCIOLOGICAL COMMENTARY.[50]

Adventures in Science premiered on Friday evening, May 6, 1938.

The renamed series marked an important break with the sedate approaches of the 1920s. Scientists suggested and invited by Science Service still appeared as guests, but a professional writer at CBS began drafting the scripts, adding dramatizations, and generally following the direction of the network executives, to the extent of completely rewriting scripts the day before they were to air. The series was described as "a presentation of Columbia's Department of Education in which Watson Davis, director of Science Service, cooperates." A script exploring Joseph Goldberger's work on identifying the cause of pellegra opens with dramatic vignettes describing his initial research ("Wish me luck. I'm going South in search of a clue"), then "swings again to Washington, D.C., headquarters of the nation's health crusade" for an interview with a scientist, and then switches back to New York for another interview. Both interviews were conducted by network announcers, not Davis.[51]

In August several episodes accentuated themes of scientific freedom. Here too CBS trivialized serious topics and introduced frivolous dramatic touches ("For twenty centuries Scientists have been adventuring . . ."):

FIRST SCIENTIST: Ahah, my friend, here we have it.

SECOND SCIENTIST: Yes, there are enough bacilli in this tube to poison the water supply of a hundred cities. The war department has better fighters in this glass than they have at the front. (*Laughs*)

FIRST SCIENTIST: The culture turned out better than I expected.

SECOND SCIENTIST (*fading*): Yes, and the culture in the incubator should yield even better results. I'm trying something new in that one.[52]

"Can the same scientist that destroys the world, save it?" the narrator wondered on this August 12, 1938, program. A list of scientific discoveries, both positive and negative, was followed by discussion of the 1933 campaign for a "declaration of independence for science and scientists," spearheaded by Robert A. Millikan and Henry Norris Russell, and then by fictionalized vignettes on the death of Socrates and the execution of Antoine Lavoisier during the French Revolution. In the section on Lavoisier, the script paid more attention to the chemist's death than to his ideas (with an unnamed voice declaring, "The Republic has no need for learned men . . . [or] for chemistry"). At the program's end, John W. Studebaker, U.S. Commissioner of Education, read a brief statement about "the attitude of the modern educator toward the world struggle to preserve the freedom of an inquiring mind."

The August 19 episode returned briefly to the older format and quality. Broadcast directly from Cambridge, England, where Davis was attending a meeting of the British Association for the Advancement of Science, the show was introduced by the head of the CBS foreign bureau. "We are speaking to you, ladies and gentlemen, from Emmanuel College, Cambridge," Edward R. Murrow's familiar voice intoned:

> During the past two days Cambridge University—for more than seven centuries one of the world's foremost centres of intellectual activity—has witnessed events that may prove extremely important to human history. For here, leading scientists of many nations have been discussing . . . how the great brotherhood of science can combat international intolerance and the suppression of truth.[53]

Murrow then listed various discoveries of "important, fundamental, new knowledge" announced at the meeting. But, he continued, technical knowledge alone will not suffice. The crucial problem is how to make the world safe for all people: "Literally . . . we need a new crusade for the preservation of tolerance and truth. This, the scientist knows."

In a cadence that Americans would come to know well during the London blitz, Murrow set the scene: "Tonight, for it is after midnight here in Cambridge, two leaders of British science are with us to bring you . . . highlights of the hopes and plans of the scientific world for preserving freedom for the future." He introduced Lord Rayleigh (president of the British Association) and Sir Richard Gregory (editor of *Nature*) and their interviewer, Watson Davis.

The subsequent discussion proceeded without sound effects or dramatizations (Davis: "Science has been blamed for inventing frightful mechanisms for use in war. Does it deserve this blame, Lord Rayleigh?"). Murrow's final remarks, offered in his characteristic style, returned listeners' attention to that summer's political events:

> Across the world today—unrest is the watchword of the hour. Unrest that touches you in every sphere of life . . . unrest that creates new problems and sharpens the anguish of old ones.

The challenge of solving these problems could be "the very sparkplug of progress," but such an effort, Murrow emphasized, requires an independent science, one free of the intolerance rising throughout the world. Murrow's work at the Institute for International Education had sensitized him to the plight of freethinking scientists and scholars in Europe.[54] What, then, could his listeners do? They could embrace freedom of expression, he declared, and they could "speak and think only that which you know is scientifically true."

The following week's episode, written and controlled by the New York staff, returned to watered-down dramatizations, and subsequent broadcasts centered on abbreviated, almost flippant interviews. On September 16 CBS staff member Paul Woodbridge introduced chemist Harold C. Urey by saying:

> We're off today on the trail of a drop of water that spread itself into a thunderstorm and washed up on the tables of research scientists a thousand new problems to face and fathom. It's the story of Heavy Water, a magic potion as fascinating as any witch's brew and the key, perhaps, to the next door of human progress.[55]

Such material did not attract many listeners. On September 30 the series was canceled. While programs like *The World Is Yours* (NBC), *New Horizons* (CBS), *Exploring Space* (CBS), *Science on the March* (NBC), and *Men against Death* (CBS) continued, Science Service was again off the air.

WHO OWNS THE MICROPHONE?

In late 1938 CBS asked Davis to resume writing and production of *Adventures in Science*, returning to the news-plus-interview format. The revived series blended attention to academic science with occasional unembarrassed promotion of industrial research and with more emphasis on the practical implications of scientific accomplishments.

Davis was now keenly aware of who controlled access to the radio audience. He wrote the scripts and invited and interviewed guests, but he knew that he must heed the network's suggestions or risk cancellation. Fisher and Sioussat kept a tight rein on the series, frequently recommending topics, guests, and approaches. In January 1939, for example, Fisher told Davis that CBS was planning extensive coverage of the upcoming conference of the American Association for School Administrators in Cleveland and wanted "to include the 'Adventures in Science' broadcast of Thursday, March 2nd, in the list of programs devoted to observance of the conference."[56] He assured Davis that there would be "several scientists attending the meetings" who could be interviewed. By the end of the month, Davis had obtained a copy of the conference program and arranged to interview Iowa researcher George D. Stoddard about "Cultivating the Child's Mind." A few weeks later, Sioussat asked Davis "to build your 'Adventures in Science' broadcast of May 18th around the annual meeting of the American Medical Association" because

CBS was planning another broadcast in conjunction with that meeting.[57] Davis replied quickly that he was working on it. After a visit from a staff member of the New York Museum of Science and Industry, Fisher suggested that Davis "use her as a contact in arranging your proposed 'science in crime' program," and when Fisher became aware of the seventy-fifth anniversary of the Columbia University School of Engineering, the result was a Davis interview with the school's dean.[58]

Other "suggestions" came from (or were prompted by) CBS advertisers. Fisher informed Davis that CBS would be "cooperating during the first two weeks of April with the Associated Grocery Manufacturers (AGM) in a 'Parade of Progress' campaign," and asked Davis to arrange interviews with scientists "from the research laboratories of large food product companies." Handwritten notes in the letter's margin indicate that Davis immediately sought suggestions of potential guests from AGM and related associations, but Fisher continued the pressure with a telegram: "WOULD APPRECIATE ANY INFORMATION RE ADVENTURES IN SCIENCE APRIL 8 DEALING WITH EARLIER SUGGESTION MADE TO YOU RE INTERVIEWING SCIENTISTS IN LABORATORIES OF GROCERY FIRMS."[59] Davis, who was on the road, asked the Washington office to locate potential guests with acceptable "marketplace" connections: "PROBABLY SAFEST GROCERY PROGRAM WOULD BE SOME REPUTABLE SCIENTISTS CONNECTED GENERAL FOODS. . . . KINDLY EXPLORE BUT KEEP IT NONCOMMERCIAL." On return to Washington, Davis informed Fisher that they were indeed "working on a food program . . . to tie in with the grocers' parade of progress . . ."[60] Lewis W. Waters, vice president of General Foods, eventually spoke on the topic of "Better Meals Tomorrow," assuring listeners, "Food scientists and the food industry are helping to build a bigger and better America of tomorrow."[61]

Davis sometimes solicited input directly from corporate public relations and advertising representatives, or entertained their suggestions, but he occasionally ran into objections related to advertising conflicts.[62] A public relations firm had brokered a broadcast featuring three clients involved in the manufacture of synthetic rubber materials, but the DuPont Company then pulled out of the arrangement because they "had their own commercial programs [*Cavalcade of America*] over which they have complete control" and were therefore "not interested in sustainings."[63]

Fisher's goal was to increase the number of listeners, not to promote science. From then until he moved to NBC in 1942, he continued to oversee the direction and content of *Adventures in Science*, demonstrating only a superficial comprehension of the topic.[64] Apparently ignorant of the coup Davis had pulled off in persuading Enrico Fermi to appear on the air in early 1939, Fisher told Davis to bump the physicist from the February 16 broadcast in favor of a program for which Fisher had arranged a "full page picture layout" in *Radio Guide*.[65] Fortunately, Davis was able to reschedule Fermi for an earlier date. (See discussion of Fermi's February 2, 1939, broadcast in chapter 9.)

The changed format, and the additional publicity supplied by CBS, brought results. At the Science Service trustees meeting in February 1939, Davis reported that the first program on January 5 had drawn seventeen hundred letters requesting more information about the topic. Closer coordination with CBS also widened the production possibilities, allowing, for example, more relevant sound effects. But Davis only held that coveted evening spot for a few months. At the start of the year, the fifteen-minute program was scheduled at 7:15 p.m. on Thursdays, then on March 25 it was moved to 6:15 p.m. Saturdays, on April 24 to 5:30 p.m. Mondays, and on September 18 to 4:30 p.m. Mondays. The series stayed in a late-afternoon, weekday slot until 1941, when it was shifted to midday Saturdays, where it remained through the 1950s. Although Davis was doubtless disappointed at not appearing in prime time, the weekend and afternoon scheduling presented science to an audience of adults who might not otherwise have tuned in to it. To Davis's credit, program topics and interviews were never reduced to simplistic summaries, and they consistently explored some of the most complex, serious issues of the times.

Perhaps the most unusual broadcast in 1939 involved a chemist who did not speak on the air at all. After the usual opening news segments on the November 13 broadcast—world weather trends ("the world is getting warmer"), innovations in home insulation materials, discovery of a large undersea mountain off Kodiak Island—Davis turned to the European conflict:

> One of the questions most often asked in connection with the war is: Why the delay in using gas warfare? The failure to use gas is puzzling to those of us who read about every man, woman, and child in warring countries of Europe fitted out with gas masks.... Why... in the opening phases of the present war... [were] poison gases not used?[66]

For the answer, he then turned to "the expert"—W. Lee Lewis, inventor during World War I of one of the deadliest respiratory irritants, chloro-vinyl-dichloro-arsine ("lewisite"). Lewis, however, never spoke on the show.

The noted chemist had been interviewed once before by Davis, in 1933, but, when invited to appear again that November, Lewis explained that he had been having vocal problems. He suggested that his wife, Myrtiela Mae Lewis, could substitute if necessary.[67] She has a "most unusual speaking voice... with exceptional enunciation," he assured Davis and, as a former chemistry student, would be conversant with the topic. Mrs. Lewis did indeed appear, reading from a script that contained frank assessments of the "usefulness" of chemical weapons and the international politics of banning all chemical weapons, including those used "to quell mobs and riots."[68] At the end of the broadcast, listeners were informed that if they mailed a postcard to Science Service, they would receive a free bulletin on "War Gases" that listed the chemical characteristics of mustard gas, chlorine gas, lewisite, and toxic smokes.

Although scientists were becoming more enthusiastic about popularization through radio, they would occasionally express hesitation at either the timing or topic choice. A. F. Blakeslee, director of the Carnegie Institution's Cold Spring Harbor Laboratory, had appeared on *Adventures in Science* in March 1938. When Davis invited him again the next year, the scientist balked, noting that while hundreds of letters arriving at the laboratory indicated that the public remained interested in his genetics research, he was not sure "if we have anything of public interest which is fundamentally new since the last broadcast."

> The making of "new species" by three laboratory methods is perhaps the matter of most general interest. This, however, is not scientifically new, although its presentation is a little different from what has been before given.[69]

Throughout January 1940 Blakeslee kept threatening to withdraw because he did not believe the current research was suitable for "general broadcast." Eventually, he arranged for his February interview ("Why People Behave Differently") to focus instead on how heredity and environment affect human personality.[70]

THUNDER IN THE DISTANCE

Historian John Burnham has pointed out that the first major survey of radio content in the 1930s did not even include science as a separate category.[71] In 1939 Watson Davis estimated that only about six and a quarter hours of educational or informational science programs were being broadcast on all of network radio every week; comparison of radio and print news sources in the late 1930s concluded that newspapers gave five times as much attention to science than did radio.[72] Although the allotted time was brief, programs

FIGURE 13. Ernest O. Lawrence, director of the University of California Radiation Laboratory, April 1939. The physicist was a frequent guest on *Adventures in Science* and on programs produced by his university's Radio Service. He was attending a National Academy of Sciences meeting in Washington, D.C. Photograph by Fremont Davis. Courtesy of Smithsonian Institution Archives.

like *Adventures in Science* were well presented and timely. The science they discussed was representative of the most interesting and significant work being done. And they frequently addressed critical social and political issues related to the economic situation, the European conflict, anxiety about America's involvement in the war, and changing attitudes toward how scientific information should be protected. Programs in 1940 and 1941 offered themes tied to national preparedness, such as a broadcast from Inventors' Hall at the New York World's Fair during which Americans were encouraged to submit useful ideas and inventions to the federal government. The scientists' campaign for large-scale public funding of research also received attention; on four programs in October and November 1941, guests like Harlow Shapley, zoologist Edwin G. Conklin, Bell Laboratories chairman Frank B. Jewett, and William J. Robbins, director of the New York Botanical Garden, described the rationale for the proposed National Science Fund.[73] In addition, the broadcasts of 1939–1941 demonstrate how mass culture was humanizing (or attempting to humanize) scientists. When Davis, for example, announced in November 1939 that Ernest O. Lawrence had just received the Nobel Prize, the news segment included not just a description of how cyclotrons work, as might have been included before, but also a brief gratuitous celebritization of the physicist ("Tall, blond and with rimless spectacles he could easily be mistaken—and has been—for one of his own students").[74]

Above all, the Science Service broadcasts projected an unrelenting tone of optimism. Scientists were always moving forward. Progress was inevitable. Europe might be in turmoil but, thanks to science, any conflict would surely be brief. On December 25, 1939, Davis opened his annual "Review of the Year" program by wishing listeners "A Merry and Scientific Christmas." He then declared that he would "draw up a sort of scientific balance sheet, to judge what has been important and significant," declaring confidently, "Long after the war of 1939 is forgotten, the splitting of the uranium atom with release of energy, hinting practical production of power from within the atom, may be listed as the year's outstanding achievement."[75] It would be, of course, far more than that. The development of atomic energy and of unimaginably destructive weapons brought changes to global stability. In altering science's political context, that discovery also altered the very freedom to communicate openly about science that Davis and others had worked so hard to nurture. War and its aftermath, which increased science's political capital, soon hobbled scientists' ability, even in pursuit of civic education, to follow Edward R. Murrow's salient advice to speak "that which you know is scientifically true."

CHAPTER NINE

——— ✳ ———

Broadcasting the Voice of the Atom

It is not always so important what you put in a paper as what you keep out. . . . It is far more difficult to decide what not to publish than it is to decide what to publish.　　　　　　　　　　　　　　　　J. F. HELLWEG, 1932[1]

ON December 13, 1941, a special half-hour version of *Adventures in Science* celebrated the dedication of Science Service's new Washington headquarters and commemorated the twentieth anniversary of the group's founding. Despite a national emergency precipitated by the Pearl Harbor attack less than a week earlier, the broadcast, speeches, and reception went on as planned. Declaration of war reinvigorated science popularization and gave it new purpose.

War, however, also brought unsettling and unprecedented limits on communication. For years, Science Service and its supporters had endeavored to *increase* the flow of scientific knowledge to the public. They had struggled to make it acceptable for the most prominent, accomplished researchers to engage in popularization, to have their work summarized on the front page, and to step into radio studios to be interviewed. In the public interest, it had been argued, science should be open and scientists' insights should be made available to all. Now, in wartime, these basic assumptions were being questioned. U.S. science and engineering were playing essential roles in military preparedness. A nation's ability to capitalize on its scientific talent and to protect its innovations might spell the difference in the impending conflict.

Governments were not the only ones imposing restrictions. Years before war was formally declared and any official censorship imposed, physicists had entered a phase of unusual reticence. Aware that the atom's power could potentially be weaponized and worried that such knowledge might be used by the repressive Nazi regime, they had begun to censor their own publications and presentations. During the 1940s, then, instead of discussing only how to facilitate communication, scientists and media organizations began to confront the reverse—whether, when, how, from whom, and for how long scientific information should be withheld.

By summer 1945, when the atomic bomb reaffirmed science's alliance with the body politic, the tone and direction of popularization had already shifted. Postwar coverage of atomic energy began in celebration and curiosity, but within months even radio—that comforting source of baseball, music, and laughter—was exploring the terrible potential of atomic warfare, the acrimonious political debate over military versus civilian control of the atom, and the chilling implications of increased scientific secrecy. Having previously been dramatized and humanized, science on the radio now became more politicized.

A HAPPY OCCASION

Since its founding, Science Service had operated in proximity to the National Academy of Sciences and the National Research Council, either in a nearby building or, from 1924 to 1941, inside the NAS headquarters on Constitution Avenue in Washington. In 1941 ten rooms were still being "made available . . . on a share-the-upkeep basis," but Watson Davis had been informed that "with the defense effort becoming more intense and the part that science was playing in it becoming greater and greater," the NAS wanted to reduce that allotment.[2] Davis and the Science Service trustees scrambled to purchase a town house near Dupont Circle and decided to hold a formal dedication ceremony and celebration in mid-December and to broadcast those proceedings live.

They invited an illustrious group of speakers: Frank Thone's friend Henry A. Wallace (now vice president of the United States); Vannevar Bush (president of the Carnegie Institution of Washington and director of the Office of Scientific Research and Development [OSRD], which would oversee the Manhattan Project); astronomer Harlow Shapley (now chairman of the Science Service board of trustees); zoologist Edwin G. Conklin (executive officer of the American Philosophical Society and president of Science Service); and Colonel Charles G. Darwin (a grandson of the evolutionist and director of the National Laboratory in England).

FIGURE 14. Broadcast of the dedication of the new Science Service headquarters building, December 13, 1941. Seated (left to right): astronomer Harlow Shapley, U.S. vice president Henry Wallace, and zoologist Edwin G. Conklin. Standing (left to right): Col. Charles G. Darwin, head of the National Laboratory in England, and Vannevar Bush, director of the U.S. Office of Scientific Research and Development. Courtesy of Smithsonian Institution Archives.

Davis estimated that probably a hundred radio stations would carry the transmission live that Saturday afternoon or record it for later broadcast. Then, on December 7, the Japanese attacked Pearl Harbor. The United States was at war on both the Atlantic and Pacific fronts. When Shapley submitted his script for the broadcast, he penciled in a postscript to Davis: "Will Wallace go through with it now? Shall we?"[3] They did.

War politicized the platform for science popularization. For over two decades, Science Service had adopted relatively nonpartisan language, walking gingerly among the controversial issues of the day, from evolution to the New Deal. E. W. Scripps had believed that science played an essential role in political decision making, that understanding of science mattered to the health of civil society, and that there was more to science than mere "knowledge for knowledge's sake." And yet many of the organization's most ardent supporters—prominent scientists, laboratory workers, science teachers—wanted to believe otherwise. As individuals, they held strong opinions about political candidates, parties, issues, and goals, but they constructed science in their public statements as aloof from such matters. Moreover, if *science itself* strived for objectivity, then *science popularized* should also be above emotionality, bias, subjectivity, and partisan politics. Of course, this image was far from reality. The debate over teaching evolution, the research directed at economic recovery during the Depression, the Nazi persecution of Jewish scientists—such episodes continually demonstrated that science resonated within its social and political context, with both good and ill consequences.

In late 1941, scientists, like all citizens, were being called to arms. Davis introduced the December 13 broadcast, saying, "In these critical days in common with all Americans we pledge our all to the arduous duty of rescuing the world from the dangers of ignorance, brute aggression and the powers of darkness." And then, in a phrase added by hand to the draft script, he declared: "Science pledges victory."[4]

Both scientists and science popularizers knew they could not remain on the sidelines, pretending that science was disconnected from current affairs. (Those informed about what would come to be known as the Manhattan Project were already aware that science was entwined in the war effort.) The speeches on that day demonstrated how thoroughly the old public image of neutrality was being repudiated. Vice President Wallace phrased the challenge in humanistic terms:

> True science, true democracy and true religion have much in common. A great prophet in the Old Testament said, "When there is no vision, the people perish." And from the New Testament comes an even greater appreciation of the unity of democracy, science and religion: "And ye shall know the truth, and the truth shall make you free."
>
> Believing that the wide dissemination of truth is fundamental to democracy and religion, we endeavor to use the privileges given us under the Bill of Rights to make education rather than propaganda freely available to our people.[5]

Wallace reminded the audience that Science Service's founders believed passionately that "truth *can* be popularized" and that "the safety of democracy in the future would depend on all of the people knowing more of the facts of science."

Vannevar Bush, introduced by Davis as a "soldier of science," continued this theme (and rhetoric) in remarks drafted for him by Frank Thone. The "blasts of enemy bombs last Sunday" had "rudely shaken" scientists' hope for the world remaining in peace, so scientists were joining the fight: "Tools in science's armory for the improvement of weapons—and they are many and powerful—will be used to the utmost for the defense of this country, for the aid of our friends, for the destruction of our enemies." Throughout 1941, Bush had been engaged in a campaign to establish and endow some sort of national research agency. His remarks included this (anticipatory) plea for lavish reconstruction of postwar science:

> What should we expect of science in the twenty years, the hundred years, that are to follow the peace that will some day come? Men of science will spontaneously and individually do as they have always done—spend their efforts, as before, for the benefit of mankind without question or thought of nationalism.[6]

Nationalism was, in fact, already on the agenda. Science journalists and popularizers were entering a decade in which their long-standing assumptions about the openness of science, the flow of its knowledge to the public, and its internationalism would be tested repeatedly in the political arena.

SPEAKING (OR NOT) OF ATOMS

To a certain extent, the popularization of science continued throughout World War II much as before. The amount of science on radio did not noticeably decline, although many series adapted wartime themes or imposed patriotic frames around their scripts. In June 1942 NBC, in cooperation with a coalition of professional societies, produced *The Engineer at War*, describing how engineering contributed to an understanding of blackouts, bombs, and structural damage. A July 1942 episode of *Adventures in Science* featured the winners of the Science Talent Search, reading from essays on the topic "How Science Can Help to Win the War." Chicago high school student Hugo Korn described his idea for a device that would enable more effective nighttime bombing raids and sixteen-year-old Beatrice Meirowitz of New York City told about the "potentialities" of uranium-235, which she called "a tool of both wonderful and frightful possibilities."[7] Other *Adventures in Science* broadcasts showcased military officers and government officials discussing scientific and engineering aspects of the war effort, Red Cross

FIGURE 15. Science Service senior biology editor Frank Thone with high school students in the Science Talent Search competition, on *Adventures in Science*, February 17, 1945. Winners of the annual contest were regular guests on the series. Courtesy of Smithsonian Institution Archives.

officials describing their blood program, nutritionists advising how to accommodate family menus to rationing, and an assistant director of the FBI describing how scientific techniques helped to combat "the spy, the saboteur and the criminal in the underworld." The American Medical Association successively retitled its programs to convey a dynamic vision of medicine's contribution to war (and peace)—*Doctors at Work* (1940–1942), *Doctors at War* (1942–1944), *Doctors Look Ahead* (1945), and *Doctors at Home* (1945–1946).

Such positive presentations gave little hint that popularization was already constrained, nor did they reveal the extent of voluntary censorship. Until December 1941 the political climate for scientific communication in the United States had favored openness. The federal government had placed few

FIGURE 16. Watson Davis, director of Science Service, with Hugh H. Clegg, assistant director of the Federal Bureau of Investigation (FBI), in a publicity photograph for *Adventures in Science*. On the February 10, 1945, broadcast, Clegg discussed "Science and Crime Detection," explaining, among other things, how FBI agents were trained to make plaster casts. Photograph by Fremont Davis. Courtesy of Smithsonian Institution Archives.

formal restrictions on civilian scientists' freedom to communicate, either among themselves or to the public, in print or via radio, a policy that reflected the nation's long-standing commitment to freedom of expression, and an attitude that had worked to the advantage of popularizers.[8] Within science journalism, the biggest barrier to open communication had usually been researchers' reluctance to have results discussed in the press before publication in a scientific journal.

Atom smashers, fission, uranium, and the possibility of an atomic bomb had in fact long provided intriguing, colorful subjects for writers and broadcasters. Science Service's news bulletins and *Daily Mail Report* service, for example, traced the work of Marie and Pierre Curie in the 1920s, and when Frédéric and Irène Joliot-Curie produced artificial radioactivity, readers were told about similar work in laboratories throughout the world.[9] Likewise, the cyclotron work of E. O. Lawrence at the University of California ("atomic bullet streams aimed at the secrets of matter's constitution") was described extensively in print and on the radio.[10] In early 1934 Science Service trumpeted that it supplied its clients the most accurate information about "essential progressive steps in the experimental exploration of matter's secrets."[11] As physicists suggested that the seemingly fanciful imaginings of novelists might actually come true—that enormous sources of power might be unlocked from the heart of the atom—science began to exhibit a "most curious mélange" of communication and silence, a pattern of open communication and "concomitant suppression."[12]

In January 1939 one of the newest Nobel laureates, Enrico Fermi, arrived in the United States. Excitement swirled throughout the scientific community. Within a few days, Davis had arranged to speak with Fermi on the phone, obtaining information for news articles and persuading him to be interviewed on the air.[13] Gossip among scientists—soon reported in the newspapers—held that German scientists Otto Hahn and Lise Meitner had achieved nuclear fission. In January, in advance of an international gathering of theoretical physicists in Washington, some participants obtained a copy of Hahn's paper published in *Die Naturwissenschaften* and attempted to confirm the result.[14] "Convinced of the reality of the energy release from uranium," Davis wrote in a January 30 news report, "there will be a great rush to complete science's current mystery problem."[15] Speculative headlines like "Power of New Atomic Blast Greatest Achieved on Earth" filled the press after the January conference, with many stories emphasizing the possibility of creating an explosive chain reaction.[16]

In introducing the February 2 broadcast with Fermi, Davis emphasized that "the world may be on the brink of the release of atomic power."[17] But physics had to wait while Davis first read the usual brief summaries of other scientific work, such as invention of the klystron radio tube, development of a faster-working blood-clotting substance, and the U.S. Biological Survey's "burrow-to-burrow canvas of the groundhog population" (February 2 was Groundhog Day). Then Fermi summarized the recent fission work and the preliminary calculations of how much energy might be released using these new techniques.

Politics had begun to color every decision physicists made, and international tensions had forced them to consider more carefully every public

FIGURE 17. General Electric Company engineers "broadcast the atom" over WGY-Schenectady, 1929. The accompanying press release described how "L. A. Hawkins . . . is busy broadcasting the smallest voice in the world—that of an atom. An atom can make itself heard in this Geiger counter which amplified the tiny change in electric potential the atom caused so that it may be broadcast over a radio." Public fascination with advances in physics, and with the technology of radio, was high throughout the 1920s and 1930s. Courtesy of Smithsonian Institution Archives.

statement. In closing, Fermi attempted to frame Hahn's research as just another attempt to gain "a better understanding of the intimate structure of matter," with no assurance of any "practical outcome." What should be celebrated, he added, was the cooperative spirit in which they were all working: "In a world that is sadly divided by material interests, it is comforting to see that beyond all barriers of nationality scientists find a common aim in the disinterested search for truth."[18]

Within a few short months, such dreams of cooperation were fading. Hahn's work had prompted Leo Szilard, Fermi, and others to question the wisdom of publishing their work in the open literature.[19] The physicists began to consider various options, given that neither the American nor British

FIGURE 18. Physicists Gregory Breit, Enrico Fermi, and George Gamow at the George Washington University Conference on Theoretical Physics, January 1939. A few days later, Fermi (center) appeared on *Adventures in Science* to discuss discoveries announced at the conference. Breit (left) is credited with having spearheaded the voluntary censorship among scientists conducting research on nuclear fission. Photograph by Fremont Davis. Courtesy of Smithsonian Institution Archives.

governments had become officially involved. Stopping research (and publication) altogether seemed unthinkable—the work was too exciting and important. Nevertheless, the European situation was deteriorating daily. The thought of such power in the hands of the Nazis was terrifying, especially to European émigrés now in the United States. Perhaps some type of voluntary restraint might work, whereby papers could be circulated privately among a select group of physicists.

The irrepressible scientific ego, the intellectual's desire for priority and credit, proved to be too strong. After a French team led by Frédéric Joliot-

Curie ignored pleas for caution and proceeded with publication in the British journal *Nature*, the floodgates opened and over one hundred papers on fission and related topics appeared in scientific journals during 1939 alone.[20] Science journalists continued to describe the work to the public (e.g., the May 29, 1939, interview on *Adventures in Science* of Westinghouse physicist A. Allan Bates, who described his company's newest "atom smasher").

During 1939 *Adventures in Science* was entering its new production phase, and the excitement within physics attracted the attention of CBS. Prompted by Sterling Fisher, the CBS Pacific Network made special arrangements in April to broadcast the first demonstration of Lawrence's latest cyclotron at the University of California, incorporating a West Coast interview of Lawrence within the *Adventures in Science* series. Fisher seems to have been especially excited about the prospective "atom smashing" broadcast—not because he grasped its scientific importance but because he was convinced it could yield "a broadcast of explanation, action and resulting sound" and entice *Radio Guide* to give the show, and the network, valuable publicity.[21] Davis agreed, knowing that Lawrence was a reliable speaker and that Science Service could coordinate arrangements directly with Hale Sparks, the university's radio administrator, who would introduce the physicist.[22] Science, of course, marched to its own drummer. The cyclotron was not ready in time for the planned sound-effect demonstration on April 15, so Sparks interviewed Lawrence instead, asking him to describe how the new cyclotron would differ from the current one.

Print and radio journalists continued to discuss atomic physics through spring 1940. Although they undoubtedly did not receive full cooperation from physicists like Rabi who advocated self-censorship, the journalists were not being pressured to exercise similar restraint. In the news section opening the May 9, 1940, *Adventures in Science*, Davis led with information about U-235, a new "atomic explosive." "Hope of atomic power was revived" by recent discoveries in Germany, he explained to listeners; since then, a dozen or so laboratories had been engaged feverishly, using the "tools of science . . . to concentrate uranium," and many had confirmed the theoretical predictions with "chain explosions of uranium atoms."[23]

That June physicists in the United States took action to censor their own communications, led by a member of the Science Service extended family. Marjorie MacDill Breit had worked for Science Service during the mid-1920s and continued to contribute articles after her marriage in 1927. Her husband, University of Wisconsin physicist Gregory Breit, is credited with having persuaded major U.S. journal editors to forward all manuscripts on fission to a review committee established by the National Academy of Sciences, which then determined whether publication should be delayed.[24] In July the National Research Council set up an Advisory Committee on Scientific Publication, chaired by Luther Eisenhart, which recommended that "reference committees" also be established within each research field to advise authors and editors on whether publication of particular research results should be postponed for the sake of national defense (with a procedure for later acknowledging priority and awarding credit).[25]

Although this advisory work was eventually coordinated with that of the federal government's National Defense Research Committee (NDRC), the scientists had taken the initiative and effectively constructed their own information control system, separate from official government censorship policies.[26] The Eisenhart Committee's July 1940 document "Publication of Scientific Work under Emergency Conditions" was not even shared with major science journalists until later that year because the committee regarded such action as tantamount to drawing "publicity" to its work.[27] Davis attempted to point out that the committee's secrecy was counterproductive. Both journalists and the public deserved to be informed of the extent of any voluntary censorship, he argued: "Because publication to the scientific world and to the public is an essential part of the research process, it should be known widely and clearly ... what general restrictions there are on publication however they may be created."[28]

A (NOT SO) BRAVE NEW WORLD: GOVERNMENT CENSORSHIP

After declaration of war in December 1941, censorship was extended on a voluntary basis to all publications, films, and broadcasts in the United States.[29] Implementation of formal rules proceeded slowly. The newly established U.S. Office on Censorship focused on protecting information about troop movements and defense facilities rather than scientific research. The office was not even informed about the Manhattan Project and the nature of its work until 1943.[30]

Government censors preferred to apply extralegal pressure. No one wanted to test the First Amendment or alienate publishers by sending editors or broadcasters to jail because the cooperation of technical publishers (whose periodicals would be most immediately affected by any restrictions) was essential for the massive programs being created to train military personnel, civilian engineers, and factory workers.[31] Blanket censorship was also out of the question, yet targeted controls required trained censors with sophisticated understanding of the potential importance of each piece of information, with the ability to distinguish between textbook science and military secret.

The government's first wartime code of practices, issued in January 1942, mentioned "new or secret military weapons" and "experiments" without further clarification of the latter term, an omission of considerable inconvenience to science journalists. The code was later amended to explain that nothing should be printed or broadcast about experimental work on such things as "atom smashing" or "atom splitting" or on elements like uranium and polonium, which were all topics that had been previously discussed in the scientific and popular literature. Patrick Washburn and other historians have shown that the press referred frequently to the potential of an atomic bomb and to other general aspects of atomic physics research throughout the war's duration, while technical details (as well as the nature of the work in Oak Ridge and Los Alamos) were not disclosed, presumably because science journalists exercised knowledgeable voluntary restraint.[32] When one *Science News Letter* subscriber wrote in December 1942 to complain about the publication's physics coverage, Davis explained that they would "like to write more about Uranium isotopes and atomic power, etc.," as requested, "but it is not possible to do this because of the secrecy connected with our war effort."[33]

The abrupt restrictions on material that had once been so freely available for public discussion created understandable confusion.[34] Sidney Kirkpatrick, editor of *Chemical & Metallurgical Engineering*, had, for example, discussed various uses for magnesium when he appeared on an *Adventures in Science* broadcast in 1939. His return appearance in early December 1941, to discuss the marine mining of magnesium, therefore seemed routine. The program had been coordinated with Dow Chemical Company and coincided with publication of "Magnesium from the Sea," Kirkpatrick's article about a Dow project. Like almost every other editor of a technical or engineering magazine, then or since, Kirkpatrick enjoyed comfortable relations with commercial firms like Dow.[35] The achievements of industry engineers and scientists filled the magazine's pages, and corporate advertising helped to support publication.

The onset of war heightened concern about protecting defense plant installations and military bases, however, and also increased sensitivity to the potential espionage value of technical information. In late December, a few weeks after the broadcast, Kirkpatrick received a letter from Dow's advertising manager, George Welles, stating that, now that the nation was at war, the company was voluntarily suspending all publicity about its defense plants and defense-related work: "From this time on . . . no one at Dow will be allowed to print, read in public, or even give out reprints of such suspended articles."[36]

Dow's new policy did not just extend to the present and future. Welles also asked the editor to return all aerial photographs of Dow factories within the magazine's files and to alter the past ("to quietly recall as many reprints of your article as you possibly can and destroy them, and to destroy any line cuts made to illustrate the magazine article" about magnesium mining).[37] This episode represented one of many assaults on press freedom experienced by technical publications as the scientific and engineering communities realigned the boundaries between openness and secrecy. Kirkpatrick responded with good-humored objection to the impractical request:

> It is indeed drastic censorship which your company has voluntarily assumed. . . .
> We thought we here were going a bit farther than is necessary under the laws,
> or under the implications of the rulings of the Office of Censorship, but we
> have not, as yet anyway, discontinued the distribution of back numbers of our
> magazine . . . or other publications that have had wide circulation.[38]

Kirkpatrick pointed out that, at the conclusion of the December 6 broadcast, with Dow's approval, listeners had been offered a reprint of Kirkpatrick's article and that Science Service was already processing these requests.

It was a revealing episode for inhabitants of the heretofore cozy world of science and engineering journalism. They were now being forced to balance the mandate of informing the public through a free press, and the long-standing commercial practices of oiling the wheels of commerce through good publicity, with fears of espionage that might assist enemy technical capability. Kirkpatrick added:

> I hope you won't misunderstand me, George, when I say that I personally think
> that your company has gone a little further than was necessary to conform
> with the spirit of the law. Our article and photographs had been cleared by the
> War Department. Furthermore, our publication has already given it such wide
> distribution that anyone who could possibly benefit from it has certainly done
> so by this time.[39]

And, he did not need to add, the research had already been discussed on the radio and heard by thousands of people.

By the end of 1942 even the most conservative professional publications were accommodating editorial procedures and publicity practices to government pressure. The *Journal of the American Medical Association* (*JAMA*), for example, had for many years mailed embargoed copies of its page proofs to select science journalists in advance of the official publication date.[40] Medical writers like Jane Stafford used the advance copies to prepare stories—perhaps adding information from interviews with the researchers—that were in turn mailed to Science Service clients for publication coinciding with the journal issue's cover date. The embargoed copies had proved to be an effective tool in medicine's campaign to maintain a good image, and the positive news coverage had enhanced the journal's cachet among professionals.

During the war, major medical journals continued to publish, but the advance page proofs were sometimes delayed by the censorship process, even though editors attempted to clear stories and topics well in advance of production. In one instance, Science Service was instructed to return its advance copy of *JAMA* because the Office of Censorship had decided that a single brief news article ("Use of DDT as a Mosquito Larvicide") must be deleted from the issue. Fortunately for historians, Stafford simply marked her own draft article "Do Not Use" and stuck the manuscript and the embargoed issue in her files.[41] Other internal memoranda and correspondence indicate that Science Service journalists negotiated frequently with military censors on sensitive material relating to medical advances as well as on the predictable areas of aeronautics, chemistry, and physics. These restrictions inevitably affected the news and topics discussed on the radio program as well.

Export controls, imposed on material mailed to addresses outside the United States, further influenced the content of scientific journals and publications available to Americans at home. To avoid the expensive option of publishing distinct domestic and foreign editions of *Science News Letter*, Davis simply eliminated altogether any text that might trigger such restrictions, a pragmatic economic decision that effectively censored what U.S. subscribers could read. Although reporters, journal editors, and other popularizers tried to work around such problems, perceptions of the value of technical information were changing, with consequences for the extent and quality of scientific information available to the public. Davis found the emerging situation troubling. Most people, he wrote, had no awareness of the extent of such controls:

The fundamental question . . . is whether the world at large and the scientific world has the same knowledge of the limitations being placed voluntarily upon scientific publications that it has of the limitations placed upon general publication by the press, through the code of censorship.[42]

By late 1942, science journalists' routine sources of information were shrinking. Many scientific meetings were canceled, ostensibly because of transportation difficulties (e.g., gas rationing, restrictions on nonmilitary travel) but also because of government restrictions on what could be discussed in public. Only a third of the usual number of papers were read at the American Physical Society meeting in 1942; certain topics were omitted altogether from American Chemical Society conferences.[43]

Even when censors did not intervene directly, scientists sometimes exercised self-restraint. Bacteriologist Selman A. Waksman had appeared many times on *Adventures in Science* when Davis asked him to be a guest once again in October 1942. Waksman at first declined, saying:

> There is a great deal of work being done at the present time on the antibacterial substances, some of which must be kept confidential during the present emergency. I would hardly be at liberty to discuss some of the more important aspects before a radio audience. These points may become clarified within 4–6 months. I would then be at liberty, and only too happy, to participate in one of your programs.[44]

Later the same day, Waksman reconsidered. Although discussion of specific applications of his university research would be off-limits, he offered to speak on the more general topic of "germs or microbes of the soil and their importance in the life of man."[45] Davis replied that substances like gramicidin and penicillin were being discussed openly (the September 5 broadcast, for example, had mentioned fumigacin and clavacin). Waksman again demurred, although his letters hint that he was also concerned about raising false hopes by discussing the substance publicly before all tests were complete. Davis, ever the diplomat, agreed in principle, but reminded Waksman of public education's higher purposes: "The excuse for talking about the soil and its bacteria is really the medical applications of these new substances, and we ought to make the radio audience just as interested in what is going on as the scientists are themselves."[46]

Radar was another topic that had been discussed openly in the scientific and engineering literature for years. During spring 1942, a flurry of magazine advertisements had even promoted the related products of various radio-electronic manufacturing firms.[47] The Office of Censorship became quite concerned about discussion of radar and in July issued guidelines for editors and writers:

The fact of prior publication should not be used to cover added description, discussion, and deduction, or to support a theory or draw a conclusion. Radar is a secret weapon within the meaning of the Code. Editors and broadcasters are especially requested to be alert to every mention of radar and military electronic devices . . . and to submit all material on the subject . . . to the Office of Censorship for review in advance of publication or broadcast.[48]

Past discussion of science, even on the radio, would not be allowed to inhibit future censorship. A word could be declared secret long after it was said.

By 1945 the censor's heavy hand even reached into science fiction. That April the comic strip character Superman was shown being given a tour of a "cyclotron—popularly known as an 'atom smasher'" and being warned that the machine could bombard him with electrons at tremendous speeds and voltage.[49] The Office of Censorship complained to the comic book's distributing syndicate that this dialogue came "too close to reality," so the plot line was changed, even though, as one writer notes, "a cyclotron cannot accelerate electrons, and the device pictured was not a cyclotron"—not to mention the fact that the topic had been discussed in far more technical detail in print and on the radio for years.[50]

RESHAPING THE POWER

Physicists, biologists, and chemists may have been hard at work behind the scenes improving the world (and attempting to win the war) but science still failed to excite the broadcasting executives who controlled scheduling. Throughout the 1940s, coverage of sporting events frequently preempted science broadcasts. The CBS network suspended broadcasts of *Adventures in Science* every fall, for example, to provide time for football. News of yet another three-month suspension in 1943 led Davis once again on the (unsuccessful) search for commercial sponsorship. When Harlow Shapley complained that educational programs were being "sacrificed to the commercialism of the moment," the director of the CBS education office, Lyman Bryson, merely replied that he too regretted that "the problem of public service is one which the radio industry has not solved."[51] As a CBS executive emphasized, "no science program could get a very high rating and that is what advertisers buy."[52]

Then, in August 1945, science claimed bigger headlines than any World Series game or college football rivalry.

To accommodate last-minute changes in scripts (and, with wartime rationing, to save paper), Davis and Thone would often get out scissors and gluepot. Eventually, the final script, as delivered, would be retyped for the record and mimeographed for mailing to listeners. In the archives today, the crumbling pages of those "cut-and-paste" drafts preserve their rhetorical evolution. During one week, events not only physically altered a script—they reshaped American science.

Harlow Shapley, the guest for August 11, 1945, had planned to focus on the state of Soviet science. In mid-July, however, OSRD had released *Science—The Endless Frontier*, outlining a national strategy for managing postwar science, and Vannevar Bush asked Shapley and other scientists to assist in employing "all our skill in publicity for the enterprise."[53] On July 20 Shapley suggested that the broadcast might focus entirely on the OSRD report.

Because Davis, engaged in various wartime activities, was traveling in Mexico, Thone was handling the radio programs. Our listeners, Thone replied to Shapley, "are more interested in news from Russia than they are in plans for their own future welfare," so he offered a compromise. The astronomer might first outline advances in Soviet science and then (ably predicting the Cold War) discuss "what we'll need to do if we want to keep up with the Ivanskys."[54] The broadcast would open as usual with a quick summary of "science news of the week." Thone began cobbling together short versions of stories that Science Service had mailed to newspaper clients the previous week. Thone, whose own scientific interests ran toward botany and entomology, had also drafted a paragraph about how DDT affects mosquito larvae.

On the afternoon of August 6, the news that an atomic bomb had been dropped on Hiroshima sent the Science Service staff into high gear. That week, they wrote background stories for wire clients, published a special issue of *Science News Letter* devoted to the bomb, and started working with Simon & Schuster on a "quickie atom book." "The atomic bomb just about blew our roof off when it popped," Thone later reported to Davis.[55] Thone changed the August 11 script twice, following both the Hiroshima and Nagasaki explosions, each time typing out the revised paragraph and pasting the new slip of paper on the draft. "It's not only the most important science news of the week," he wrote, but also "of the whole lifetime of every person within radio range of my voice. We are on the threshold of a revolution in the use of power." To "say anything more," after all that had already been said that week, Thone knew, would be "anticlimactic," so his brief introductory remarks, before switching to Shapley in Boston, focused on the "pooled interests," the contribution of scientific refugees from Nazi Germany, and the scientific community's willingness to make "use of human talents without regard to the race or creed of those who possessed them."[56] This was a theme of internationalism and interdisciplinary cooperation that Science Service would emphasize repeatedly throughout the next decade.

The lifting of security restrictions meant a host of exciting topics and guests were suddenly available to radio. The following week, Thone's broadcast centered on another topic that would dominate postwar popular science—chemistry and the environment. His guest, G. J. Haeussler of the U.S. Department of Agriculture, identified "weapons for the next campaign"—a "war of extermination" against the insects that cause human disease and destroy essential food crops. Censorship had inhibited discussion of certain chemical exterminating products that the military had used effectively to control lice that spread typhus and mosquitoes that spread typhoid. Now the War Production Board had announced the availability, for civilian and agricultural use, of limited amounts of "a weapon that promises to be almost comparable in its way to the atomic bomb in human warfare"—DDT.

PUBLIC RELATIONS

Throughout the United States, families and communities attempted to return to civilian normalcy. In the bureaucracy-soaked capital, planning for science careened forward at a frantic pace. As Thone wrote to one old friend, "Life here in Washington goes on almost as if no atomic bomb had ever been exploded."[57] Enormous amounts of public and private money were flowing to researchers and their institutions, pushing toward the "endless frontier" envisioned in the Bush report. Now that the scientific establishment had become comfortable with the idea of public outreach, they began to use popularization more strategically in the next campaign—for political support.

The scientific community in America did not enter this battle unprepared. During the 1930s a group of creative, intelligent, highly motivated scientists had expanded their political activities and influence. It might be easy to look at photographs from Washington events in 1938 and 1939 and interpret the faces of E. O. Lawrence or Vannevar Bush as boyish or unsophisticated, but these were not the absentminded professors celebrated in cartoons, or the white-coated, wild-eyed mad scientists of Hollywood movies. They were congenial, neat, occasionally dapper. As they navigated the corridors of power and learned how to manipulate the political system, their shoulders were straight and their demeanor correct. Prewar efforts to establish a national research foundation may not have been successful, but the political connections the scientists made during that campaign paved the way for eventual success. Even before fighting had ended in European farm fields and on Pacific beaches, the drafting of *Science—The Endless Frontier*, which emphasized science for civilian benefit, declared the scientific community's resolve to acquire government research funding on an unprecedented scale.

One particular series of radio talks exemplified both the typical approach to gathering public support and the sophistication and self-assurance the scientists now brought to such efforts. The U.S. Rubber Company had been sponsoring live broadcasts of the New York Philharmonic-Symphony Orchestra since 1943 when the company's advertising agency suggested that, beginning in January 1945, the fifteen-minute intermission talks should focus on science and "look forward into the future" ("The time [has] clearly come when everyone ought to have a broader and more authentic understanding of what science is and how it operates"). A prestigious advisory committee, chaired by Warren Weaver of the Rockefeller Foundation, was appointed to guide selection of speakers and topics.[58] The committee included Harlow Shapley, physicist Frank B. Jewett (current president of the National Academy of Sciences), Carnegie Institution embryologist George W. Corner, chemist Wendell M. Stanley of the Rockefeller Institute for Medical Research, and historian Douglas Southall Freeman. Although the group initially approached the project with "incredulous skepticism," concerned that the product might become "jazzed-up" science rather than "real science, dignified and authentic," those fears proved groundless.[59] In 1945 and 1946 concert audiences heard Arthur Holly Compton, Irving Langmuir, Isaiah Bowman, and E. P. Hubble. Edwin H. Land discussed "Polarized Light," Igor Sikorsky described "Direct Lift Aircraft," Selman A. Waksman described streptomycin research, and Robert M. Yerkes talked about chimpanzees as the "servants of science." Other talks covered DDT, hybrid corn, and heredity. The December 1945 talks centered on atomic energy: Hans Bethe ("Within the Atom"), Harold C. Urey ("Isotopes in Atomic Research"), James Franck ("Medical Benefits from Atomic Energy"), and J. Robert Oppenheimer ("The Atomic Age").

The advisory committee recognized that such broadcasts furnished a special opportunity to communicate science to a national audience, so Shapley asked Watson Davis to listen to the first few broadcasts and provide an independent evaluation for the committee. Davis thought that they aimed at a "rather low emotional level" and were not likely to "bring in the mail."[60] On the latter point, he was wrong. Beginning in May 1945, listeners were invited to write for copies of the talks. Eventually over 250,000 people wrote in, often enclosing elaborate letters praising the programs.[61]

Like the talks arranged by Austin Clark during the 1920s, the U.S. Rubber series ostensibly allowed scientists to "speak with their own voices, describe their own work, tell the stories that they themselves considered significant."[62] The intention was to give simple, straightforward, first-person accounts of research. The stories, however, were not exclusively neutral, dispassionate, or apolitical. Chemist Irving Langmuir included a plea for the freedom of scientific research; other speakers praised science's contributions to the war effort and described how research would now stimulate national economic progress. There were also expressions of concern about how (and by whom) postwar scientific research would be directed. Raymond B. Fosdick, president of the Rockefeller Foundation, declared that the recent "disavowal of concern for the social consequences of science" by many of the Manhattan Project scientists might seem at first to be "callous and irresponsible," but

> we may be facing a situation where no other answer is realistic or possible. To ask the scientist to foresee the use—good or evil of the use—to which the results may be put is doubtless beyond the realm of the attainable. Almost any discovery can be used for either social or antisocial purposes.[63]

Science, Fosdick continued, "merely reflects the social forces by which it is surrounded": "When there is peace, science is constructive; when there is war, science is perverted to destructive ends."

BIKINI

Controlling the atom politically meant controlling information about atomic science. In the postwar years, two conflicting approaches to such scientific information emerged: defensive (censorship and classification) and proactive (national and international public relations activities and education).

Even after official restrictions on the press were lifted, the U.S. War Department declared that controls would remain in place on certain sources and types of information. Everyone connected with the Manhattan Project, for example, must "continue to comply with present security regulations," all citizens (including the press) must keep discussion of the subject of atomic energy "within the limits of information disclosed in official releases," and discussion of all technical aspects of the atomic bomb's design and detonation would be restricted.[64] Passage of the Atomic Energy Act of 1946 further tightened federal authority over information relating to atomic energy, from scientific research to engineering design. With this legislation, legal analyst

Harold P. Green points out, U.S. officials were for the first time in the nation's history explicitly authorized to engage in broad-scale control of the
dissemination of technical information in private civilian contexts as well
as within government.[65]

Ironically, at the same time that governments throughout the world were
restricting technical information, they were also eagerly promoting their scientific achievements. All nations (and eventually the agencies of the United
Nations) began to use popular science outlets more aggressively in public
relations campaigns. Within the U.S. Atomic Energy Commission, the same
public information office that advised journalists, publishers, and broadcasters on whether their articles or scripts contained restricted data was also
tasked with encouraging them to publicize the agency's scientific work.[66]

One of the most spectacular public relations events in the immediate postwar period was Operation Crossroads, in 1946, an attempt to study the impact
of an atomic bomb in naval conflict and to impress potential enemies with
the weapon's power.[67] Over ninety vessels, including decommissioned bat-
tleships and submarines, were anchored at Bikini Atoll, a remote cluster of
islands in the Pacific Ocean, at varying distances from the test blasts.

Rather than conducting the tests in secret, the U.S. Navy encouraged major news coverage by print and broadcast journalists from around the world
and also orchestrated considerable media attention during the preceding
months. On one April broadcast Davis interviewed a top Navy physicist and
the officer in charge of technical operations. In late May CBS produced an
hour-long, late evening discussion of Operation Crossroads, broadcast live
from the Library of Congress's Coolidge Auditorium.[68] After a brief statement
from the task commander, who was on board the U.S.S. *Mount McKinley*,
then heading to the Marshall Islands, moderator Robert Trout introduced a
wide array of guests—from Albert Einstein and Harold C. Urey to ordinary
citizens, politicians, judges, and public officials like former vice president
Henry A. Wallace, who was now secretary of commerce. Urey engaged in
discussion with a Minneapolis housewife about the effectiveness of keeping
atomic secrets. Other participants in the program explored the arms race, international cooperation, the effectiveness of arms treaties, and constructive
uses for atomic energy.

Eager for Science Service to participate in the Bikini coverage, Davis arranged for Harlow Shapley to attend the first test (code-named "Able") as
the organization's official correspondent and to broadcast from the site. When
that test was rescheduled, the new July date conflicted with the astronomer's
schedule, so Thone went instead, along with prominent science journalists

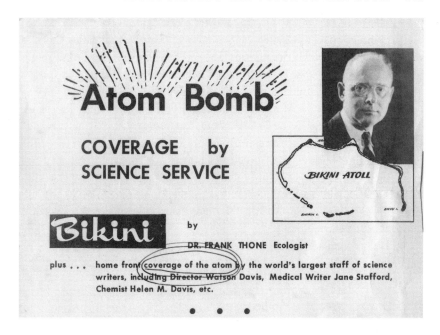

FIGURE 19. Advertisement for Science Service coverage of Operation Crossroads, 1946. The U.S. military encouraged intensive science news coverage of the atomic bomb tests at Bikini Atoll in the Pacific Ocean. Courtesy of Smithsonian Institution Archives.

like Robert D. Potter (now working at *American Weekly*) and David Dietz. Although the first broadcast announcements and reports from close to the blast were pooled, the Navy provided a ship with transmitters and a re-broadcast station on Kwajelein Island. They promised that reporters would be allowed to broadcast and write freely on what they could see.

The world was then in a brief, almost giddy period of atomic acceptance. As historian Paul Boyer documents so well, most Americans still perceived the bomb as the trick that had ended the war and not yet as the terrible weapon whose effects would be described graphically that summer in John Hersey's *Hiroshima*.[69]

Thone sailed from California aboard the press ship S.S. *Appalachian*. He and other representatives of the fourth estate, the press, who were associated with the newly emerging fifth estate, science, had new status in the world of journalism.[70] Their colleagues at newspapers and radio stations had raced to them for explanations of the atomic bomb. Some, like Potter and Dietz, had traced the birth of modern physics in their stories since the 1920s, and had met and interviewed Lawrence, Oppenheimer, Urey, and others involved in the Manhattan Project. The military knew that co-opting this group might bring favorable coverage for the test and help in the political fights to come.

Journalists are a cynical lot, though receptive to dark humor and open to being entertained. That April *Industrial and Engineering Chemistry* had sponsored a cocktail party for reporters that featured an "atomic bomb cocktail" ("While its ingredients are a military secret, its effects are well-known," the journal's editor boasted to Davis).[71] A stopover for the press ship in Honolulu included extensive wining and dining arranged by the "Operation Crossroads Hawaii Entertainment Committee," and fake newspaper editions offered a welcoming "Aloha" to the visiting journalistic "A-Bombers" enroute to Bikini.

The light mood persisted at the test site. One of Science Service's stringers, University of California News Service journalist Daniel Wilkes, who covered the second ("Baker") round of Bikini tests, wrote to Davis:

> There were a lot of guys covering it like a murder or a five-alarm blaze. . . . If their copy is anything like their verbal reactions, most of them pooh-poohed the radioactivity angle of the Baker test. Except for two or three science reporters they were unable to understand it, and, being unable to see any visible results from it, were ready quickly to dismiss it as a highly overrated affair.[72]

Wilkes complained that all the relevant technical experts were, in fact, kept away from the journalists, who were confined to the S.S. *Appalachian* ("as Thone told you and you wired me [the] reporting enterprise seemed to be severely restricted").

The Bikini tests left knowledgeable observers concerned about how atomic power would be managed. Upon returning to the office, Thone told one editor that the project "seems to have been a technical success, insofar as furnishing information useful to naval architects and engineers"—and yet:

> Full and final success . . . cannot well be claimed unless what we have written, and shall write, impresses the American people with the seriousness of their present and future status as custodians of the most revolutionary source of power that man has ever been able to wrest from nature. Prometheus had a snap, compared to what we are facing now.[73]

Thone had gone ashore on Bikini Atoll with other scientists and journalists three days after the blast and he "couldn't find a leaf loose or a flower-petal awry. . . . Nature just doesn't seem to care much about our fancy firecrackers."[74] But Thone the ecologist also complained in his dispatches that, to eradicate the flies plaguing the advance teams, the military had sprayed the islands with DDT, thereby destroying any insect life that might have been studied, and affecting the food sources of native birds and fishes: "This one monkey-wrench, thrown into the atoll's ecology, sprinkles question marks all over the biological record."[75]

DOCUMENTING THE DEBATE

In those years, popular science content began to seem incomplete without attention to the social, political, economic, and ethical aspects of science—*and* to the impacts that society, politics, economics, and moral values were having on science. A broadcast or magazine article that offered only simple summaries of research results or descriptive talks about current knowledge now seemed sterile and inadequate.

Each discipline faced its own postwar challenges in communicating with the public—none with greater consequences for the world than physics. As one scientist asked Davis, could journalists indeed "cover the atom like one 'covers the waterfront' "?[76] Could such complex social, political, and moral aspects be parsed? Could this most sensational of scientific developments be discussed without sensationalism?

The harnessing of the atom's power had unleashed fierce political debate. What to do with the knowledge? With whom should that knowledge be shared? Many technical aspects of atomic energy were declared off-limits to popularizers, and it was impractical to attempt to explain the basic physics and chemistry of the atom in a ten-minute radio talk. The social and political issues, the moral debate over the use of atomic weapons, and the government restrictions on scientific freedom all provided more fertile topics for broadcast coverage, reducible to a few provocative questions.

Radio, in fact, served an important role by facilitating public debate during this period, and it assumed this role as soon as press censorship lifted. On the August 12, 1945, *University of Chicago Round Table*, Robert Hutchins asked his guests (scientist Rueben Gustavson and sociologist William F. Ogburn), "Gentlemen, is the atomic bomb good or bad for the earth?"[77] In the following months, the Association of Scientific Workers and newly formed Federation of American Scientists (FAS) actively encouraged their members to use radio in their educational campaigns and political lobbying for civilian control of atomic energy.[78] Davis arranged a November 1945 program around questions ("Can it be kept secret?" "What do you mean by secret?")

posed to former Manhattan Project scientists and FAS representatives J. L. Rosenberg, Robert Nowak, and R. N. Lyon.[79] As Davis emphasized in his introduction, "The utilization of atomic energy for war or peace has become an international question, not just simply a scientific or technological one," and one of the consequences has been "the organization of scientists who desire to express their views to the public, to Congress, and to the officials of the Government." The physicists, he explained, would assert that there was no "secret," that the "accumulation of solutions of scientific problems" were based on "scientific principles known before the war," and that similarly skilled scientists elsewhere in the world could re-create the Manhattan Project in only "two to three years."

Such circumstances should not, however, eclipse the related issue of the evolving political culture of secrecy and its potentially damaging effects on basic research. Science had long asserted the importance of free and open communication. Now scientists were being asked to remain silent for the sake of international peace or national security, and to do so in regard to what they regarded as basic knowledge. Nowak, a mechanical engineer who had worked at the Oak Ridge facility, explained:

> The experience in science and technology is that secrecy just doesn't pay for the person who keeps the secret. . . . The nature of the world cannot be a secret and the purpose of all scientific work is to acquire knowledge and pass it on to others. The best protection against others making atomic bombs to use on us is to require that scientists all over the world tell all they know. It would then become a crime not to tell everything that is known about the atom.[80]

Throughout the 1940s, network radio attempted to explore such complexities of the new atomic age through debate, discussion, and documentaries. New York City station WCMA broadcast a thirteen-week series on atomic energy, *One World or None*, sponsored by a group headed by Albert Einstein, Harold C. Urey, and Robert Oppenheimer. Urey was also a featured guest on the July 9, 1946, edition of *Frontiers of Science* on CBS. Working with FAS, the ABC network produced a four-part dramatic reading of John Hersey's *Hiroshima* soon after the book's publication in 1946.

Documentaries proved to be among the most effective vehicles for addressing the topic. Longer than news reports and centered on serious themes, the special programs exploited entertainment techniques and used professional actors to re-create events or impersonate real people, sometimes interspersing the voices of real experts. "The material is fact and the method is fiction," one network executive explained.[81] A June 30, 1946, broadcast of *Exploring the Unknown* opened with a brief history of the Manhattan Project and then launched into a "fictionalization" of an atomic war, narrated by actor Clifford Fadiman. At the end, the sponsor, Revere Copper and Brass Company, urged listeners to write for more information about competing international plans for controlling the atom.[82] That summer, CBS produced nineteen fifteen-minute programs for its *You and the Atom* documentary series, running five nights a week from July 22 through August 16. Harlow Shapley, General Leslie Groves, and a wide array of military and scientific experts explained the development of atomic energy and the political issues surrounding domestic and international control. Watson Davis appeared on the fourth broadcast with *Saturday Review* editor Norman Cousins (just returned from the Bikini test) to discuss the atomic bomb, with Davis summarizing data from the newly released bomb impact surveys. In introducing his grim account of human casualties in Japan, Davis asked, "Is it any wonder that we have difficulty keeping the atomic bomb out of our talk?"[83]

Not every program adopted a gloomy tone. *The Sunny Side of the Atom* purported in June 1947 to be "radio's first comprehensive survey of the constructive use of atomic energy," and promised to "reveal a silver lining rich in the promise of human happiness." Actress Agnes Moorhead played a "peripatetic" narrator who toured hypothetical research laboratories, physicians' offices, and medical research facilities; another fictional character assured the audience that "folks around here won't feel so bad about that atom bomb" when they learn about potential medical applications.[84] A CBS executive later described that documentary as an attempt to "counter the 'closed-mind' attitude resulting from the 'scare' approach to atomic energy," and network surveys found that 46 percent of listeners claimed to have felt "less fearful" after hearing the broadcast.[85]

Other radio documentaries like *The Fifth Horseman* and *The Quick and the Dead* framed the development of atomic energy as more threat than promise. Fred Friendly's four-part production, *The Quick and the Dead* (broadcast by NBC in July and August 1950), alternated dramatizations of the creation of the atomic and hydrogen bombs with recordings of interviews with real scientists and politicians, and with crew members of the *Enola Gay*, the B-29 that had dropped the atom bomb on Hiroshima. Comedian Bob Hope played the role of a skeptical taxpayer who asked questions to a newspaperman, who then referred them to experts (whose answers had been previously taped); other actors played the parts of scientists (Helen Hayes portrayed Lise Meitner, while Paul Lukas played Albert Einstein).[86] By 1947 Americans reportedly regarded radio as "their most trustworthy source of information about the bomb," considering the medium faster and "more reliable" than print.[87] In the first five years after the war, hundreds of radio documentaries about atomic energy—on all aspects of the political debate—were broadcast in the United States.[88]

The FAS and other scientific organizations continued to urge public debate about scientific secrecy, especially after it became known that both the United States and Soviet Union were engaged in developing hydrogen bombs with destructive potential far surpassing that of atomic weapons. "No question is more hotly discussed in scientific circles today and no problem is potentially so important to the future of science and the nation than that of scientific secrecy," Davis explained in his introduction to an April 29, 1950, *Adventures in Science* interview with FAS members Victor F. Weisskopf and Samuel K. Allison. In a spirited and provocative conversation, the nuclear physicists argued that, while there were genuine concerns about how a potential enemy might use certain knowledge, open discussion of the hydrogen bomb was essential. "By keeping some fundamental development secret," Weisskopf emphasized, "we prevent the public from understanding the political implication of a possible application, and hence we prevent a rational and intelligent discussion of the possible effects." This assumption—that intelligent discussion of science contributed to the public welfare—was one that Science Service and all participants in *Adventures in Science* had made for decades. They had attempted to convey "all that can be told with propriety and assurance," adhering to government restrictions while also pressing for openness.[89] As Allison now explained, the Cold War had made the "subject of secrecy" perplexing. Scientists knew that to suggest any single solution could be misleading:

> DAVIS: Do you mean that you're in trouble if you do, and in trouble if you don't?

> ALLISON: That's just it. No one of us can guarantee that our potential enemies won't make military use of some new result in physics, found by us and openly published in our scientific journals. On the other hand, every one of us knows that a healthy and progressive science of physics is very important for the security and welfare of our country, and that free distribution and discussion of basic information is the very life blood of our science.[90]

Such discussions exemplify how, for a brief period after the war, radio actively assisted public debate on science. Radio, Ernest L. Stebbins observed in 1945, "reaches those who can't read or won't read, those who can't get to a meeting house, lecture hall, or movie theatre. And the radio can utilize every odd moment—while you are shaving, for example. Here, indeed, is a potent agent for good or evil."[91]

INCREASING THE CAPACITY FOR LEARNING

Thone's mantra for "Science in the Post-War World"—if scientists "are given leave to go ahead by those who control the power and the purse strings, we can and will fix up a framework for the ordinary man's life that was [once] beyond the dreams of princes"—served as the theme for many postwar broadcasts.[92] Sherman Dryer's award-winning series *Exploring the Unknown*, for example, starred major Hollywood actors in dramas about scientific achievement. During the 1946-1947 season, the show was heard by between 3 and 5 percent of all U.S. households.[93] In spring 1946, CBS created a fifteen-minute series, *Frontiers of Science*, in clear imitation of *Adventures in Science*. The chatty and personable host, John Pfeiffer, described as CBS "science director," anticipated television's "breaking news" approach to popularization. "I've just received permission to release a news story that won't be announced for a few days," he told listeners. "It's about the finding of important fossil bones in a most unlikely place—a beer cellar."[94] General Electric's *Science Forum* continued on the air and increasingly explored such natural resources topics as water quality, wildlife protection, and soil conservation.

Political debate over control of atomic energy could not be ignored, however, and the issue drifted inexorably into the subtext of programs on other topics. In January 1947 Davis arranged to broadcast live from a groundbreaking conference on heredity, fossils, and evolution, held as part of Princeton University's bicentennial celebration. His guests included anthropologist G. H. R. von Koenigswald (recently returned from Java, where he had been imprisoned during the war), distinguished vertebrate paleontologist Alfred S. Romer (just about to take over directorship of the Museum of Comparative Zoology at Harvard), and population geneticist and evolutionist Theodosius Dobzhansky.

In the introduction, Davis mentioned a conferee's statement, quoted in news coverage, that if humans "allow" themselves to become extinct, then rats and mice might inherit the earth.[95] He then asked von Koenigswald to describe his famous discovery of *Homo soloensis*; Romer spoke about "evolutionary progress" among different species; and Dobzhansky explored how heredity relates to environment. The fourth guest, the irascible British geneticist J. B. S. Haldane, then brought the discussion back to the inescapable topic. "If we all kill one another it won't be the rat that will succeed us," he

FIGURE 20. Watson Davis and his guests on *Adventures in Science*, January 4, 1947, broadcast from the Princeton University Bicentennial Conference on heredity, fossils, and evolution. Left to right: vertebrate paleontologist Alfred S. Romer; Davis; geneticist and evolutionary biologist Theodosius Dobzhansky; geneticist and physiologist J.B.S. Haldane; and geologist and paleontologist G.H.R. von Koenigswald. Courtesy of Smithsonian Institution Archives.

declared. "Since you Americans got your atomic bombs some of you have been toying with the idea of universal destruction like a man feeling a hollow tooth with his tongue."[96] The world must slow down and consider its next steps wisely, he argued. "If the physicists and politicians give us the time," we can begin to "evolve away from the ape [and] towards the ideals which our noblest men and women have set us." Survival, he stressed, depended upon the ability of human beings to live together peaceably as well as on increasing the human capacity for learning. This was certainly not the type of science programming envisioned by early radio popularizers, or that promoted by educators in the 1930s, but its potential for civic enlightenment was great. In broadcasting's next phase—the decline of prime-time radio and the rise of television—such attention to the social, political, and ethical debates within and about science continued in force and, often, with considerable creativity.

CHAPTER TEN

———— ✳ ————

Illusions of Actuality

Now for the first time, by way of television, millions of people may simultaneously look over the shoulder of the scientist as he demonstrates and describes his scientific research.   LYNN POOLE, *SCIENCE VIA TELEVISION*, 1950

TO watch recordings of television science from the early 1950s is to journey back in time. The black-and-white images seem washed out and dull, the scientists stiff, awkward, unsure of where to look and when to speak. There stands seventy-two-year-old Austin H. Clark, pointer in hand, telling viewers about caterpillars and butterflies—their migratory habits, uses as food and decoration, effects on crops and gardens.[1] He drones on about the (presumably) colorful specimens arrayed in cases on a wall, occasionally cracking a wry joke—still confident, imperious, and authoritative, oblivious to the camera. All he needs is a tuxedo to complete the image he had once projected to radio listeners.

Almost thirty years before, scientists like Clark had constructed radio as an extension of the lecture hall, a space packed with receptive auditors. Logic and rhetorical precision prevailed in the scripts, speakers took time to explain the complexities and qualifications of research conclusions. Science retained an impersonal and apolitical formality. The guests of Clark and Watson Davis imagined listeners leaning forward in living rooms in Boise and Boston, eager to learn about echinoderms or Mars or tung oil or atoms and willing to wait for comprehensive explanations and essential definitions.

The new medium of television provided no such comfort zone. Television accelerated the pace of content delivery and raised audience expectations for production quality. The medium stimulated the development of new types of science popularizers, people who were professional and relaxed on camera but who were neither scientists nor science's official representatives, hosts who projected a well-mannered image of pleasant amateurism, almost as if too much sophistication might render the science suspect. Television also enhanced and regularized attention to science's social context. Personality and politics became familiar components of popular science.

During the late 1940s and early 1950s, radio was gradually displaced as Americans' primary source for entertainment and news. Television's science entrepreneurs at first attempted a delicate compromise between how they believed science should be presented and the glitz, glamour, and pace demanded by the new medium. Yes, television is "show business," they admitted, but a television *science* program must not be like "vaudeville." Yes, television should be "entertaining, have a fast pace and employ any number of amusing 'gimmicks'—but it should always stay within the realm of dignity."[2]

Competition for public attention was even fiercer than on radio. Droning, tweed-coated scientists, accompanied by a few electrical sparks or wriggling mice, had little chance when pitted in the prime-time schedule against Milton Berle's wicked grin or Arthur Godfrey's ukulele. Whenever science did succeed on television, it was through substantial investments of money and creative talent and through approaches that acknowledged science's social and political relevance, humanized its celebrities, and held science open to questioning. Science was metamorphosed into theater. With sufficient money and creative talent, explanations of theories and discoveries could be made interesting by supplementing them with films of exploding volcanoes, animation of the swirling rings of Saturn, or actors dressed as Newton and Galileo.

Television pretended to show the audience scientific "reality"—or the illusion of reality—by adapting the same visual tricks that professors had employed for generations to keep students awake in the classroom, popping eggs into milk bottles or filling beakers with fizzy chemicals.[3] By projecting images directly from microscopes and telescopes, television producers convinced viewers that they were sharing an immediate visual experience with a studio scientist. Informational science was transmuted into show business.

With painted backdrops and prop instruments, some shows attempted to re-create the laboratories in which researchers worked, while remote broadcasts promised viewers that they might glimpse researchers in action. Interviewing a scientist in his laboratory had made little sense on radio. The 1930s actuality broadcasts had relied on skilled announcers to narrate events or describe the scene. Now, film crews could slide smoothly down the halls of science and peer into laboratories to show equipment in operation; they could interview primatologists alongside their marmosets or paleontologists kneeling beside newly excavated bones.

The expense, timing, and technical difficulties of those first live broadcasts forced ever more reliance on professional intermediaries. Amateur mistakes were costly. Viewers found them tedious to watch. Few scientists seemed comfortable in the studio; their eyes would be glued to the equipment or specimens, and they made minimal emotional connection with the audience. Milton Berle, of course, looked right into the camera and made *you* laugh; his performance was for *you*. The body language of scientists who appeared on television in the 1950s too often implied indifference. And so moderators and hosts who were not scientists told jokes to soften the tedium or interrupt pontification. These telegenic science mediators eventually reshaped scientists' access to the airwaves. Within a few years (as had happened on radio), the hosts and moderators became the preferred stand-ins and interpreters, a shift that historian John Burnham suggests helped to loosen further the scientific community's control of its own public image.[4] Once again, broadcast science became "brokered" science.

In the television era, science programming survived at the network level only if it conformed to a fast-paced, irreverent, time-conscious context. Communication of scientific information via television became fragmented and dominated by entertainment values. Popular science was shuttled to the sidelines by network executives unconcerned about altruistic goals of public education.

During the 1950s and 1960s, the proportion of television's daily schedule allocated to explaining science declined at the same time that science's accomplishments were gaining ever greater relevance to society. American television viewers never demanded otherwise, and the scientific community turned back to the laboratory with only sporadic, token, and generally ineffective complaint.

Two types of programs from that early decade of television demonstrate broadcasting's rapid shift toward entertainment. The first, exemplified by *The Johns Hopkins Science Review*, adapted radio's educational approaches to a visual context. These presentations retained radio's tone of control and dignity. The second, introduced by a group of one-hour specials underwritten by the Bell Telephone System, altered the landscape of broadcast science as thoroughly as bulldozers can reshape a hillside. Nothing looked quite the same afterward. The Bell-funded programs demonstrated to the television industry that science need not be dull. They introduced exciting visual techniques for presenting science, and, perhaps most important, they raised audience expectations for popular science. The first of those specials appeared in 1956, as the Hopkins television series was sputtering through its eighth season and two years before Watson Davis produced his final radio show. The project marked an important transitional stage in science popularization overall, from an era dominated by authoritative, objective formality to one comfortable with light entertainment and emphasizing the humanization and subjective questioning of science. Political and social themes—religion, war, environmental change, nuclear proliferation—became standard. Controversy was no longer avoided. Viewers were not expected to listen in rapt entrancement, sitting straight in their chairs and taking notes as if for a test; they were invited to become engaged in the science and encouraged to ask questions.

AN EXPLODING MEDIUM

When wartime restrictions on the development and use of television technology were lifted, the broadcasting industry in the United States entered a period of unrestrained expansion. Local station managers scouted for potential programs, much as their counterparts in radio had in the early 1920s. In 1946 a mere six commercial stations were on the air and only eight thousand households had television sets. Within less than a year, Americans owned sixty thousand sets, forty-seven thousand of which were in New York City (and three thousand of those in bars). By fall 1948, more than forty stations were on the air, serving several dozen U.S. cities, and all four networks—ABC, CBS, DuMont, and NBC—offered national programming. Adoption of the technology continued at that breakneck pace. Within three years, ninety-seven commercial stations were in operation around the country and more than 3.8 million households owned sets. Another three million sets were sold between January and June 1950; by 1956 over three-quarters of all American households (almost thirty-five million) owned televisions and there were more than four hundred stations nationwide. By the end of the 1950s, 90 percent of U.S. households owned television sets.[5]

Improvements in broadcast quality, less expensive sets, and reliable intercity cable connections helped to bolster this success. Television strengthened technically and programmatically. Initially, Larry James Gianakos observes, there was a tendency to regard television primarily as a "carrier" of events and information, to be used for dissemination, not distraction.[6] But as the quality of comedy and drama presentations improved, television quickly supplanted print, radio, sports events, and the movies as Americans' prime source of entertainment.[7] Radio continued to "structure" people's days, awakening them in the morning, informing them during the daily commute, and providing background music for chores and leisure activities, but television gradually claimed the roles that radio had played in evening entertainment and in public education, including science education.[8]

Prewar efforts to communicate science via television had been uninspiring. General Electric's pioneering television station in Schenectady, New York, broadcast a program on December 13, 1940, commemorating the founding of the company's research laboratory, which opened with a dry technical demonstration of a bonding technique.[9] After the war, the DuMont network offered what was probably the first national science series, *Serving through Science*. The half-hour program, telecast weekly at 9:00 p.m., beginning June 18, 1946, had been created for the U.S. Rubber Company by its advertising agency.[10] The title of the series was, not without coincidence, U.S. Rubber's corporate slogan. Educational films produced by Encyclopaedia Britannica were interspersed with moderated discussions.[11] When viewers reportedly found the format "boring," the network added an unrelated musical segment, and when that change failed to attract larger audiences, the program was canceled in May 1947.

In the 1940s NBC began using the term "public service" to designate broadcasts that were neither entertainment nor news. The network declared that "education" was too "austere" for a category that now contained shows intended to be "amusing, entertaining, or easy to listen to."[12] The network executives knew what would attract and hold viewers' attention. "The highest ratings," CBS education director Lyman Bryson observed, "have been for a long time, and probably will continue to be, the comedians who can count their listeners in the tens of millions."[13]

CONGENIAL HOSTS

The most successful of the early shows featured uncritical perspectives on science, content directed at adults, and guest scientists interviewed by congenial hosts. As with radio, the organizers set ambitious goals. Americans had "a genuine thirst for scientific knowledge," astronomer Roy K. Marshall wrote:

> Satisfying that thirst will . . . create a friendly attitude toward science and scientists which will favor the cause of science in the future. Science needs an informed and friendly public to back it up.[14]

Marshall typified one type of science host—a telegenic scientist. Although some producers argued in the 1950s that viewers would perceive scientists as experts and not expect them to be polished performers, the medium favored those who were animated and smooth, much as Clark had succeeded because his radio voice, deep and precise, conformed to the broadcasting stereotype.[15] Marshall had had some radio experience in the 1940s hosting *Great Moments in Science* (a local Philadelphia program for science students), but did not set out to be a television performer. In January 1948 a local television station invited him to promote a "Trip to the Moon" exhibit at the Franklin Institute's Fels Planetarium. The producers persuaded Marshall to stand in front of a "mock-up of an imaginary control panel in a space ship of the future" and jiggle a little as he pretended to take off for the moon. That appearance proved so successful that the station began half-hour remote broadcasts from the planetarium (under the title *Living Science*) and then, with the institute's cooperation, created *The Nature of Things*, a weekly fifteen-minute program that premiered only a month later, on February 5, and turned Marshall into a celebrity. Broadcast nationally on NBC starting on December 13, 1948, *The Nature of Things* ran in prime time through September 1950, returned for the next few summers as a temporary replacement for entertainment shows, and was a weekend afternoon feature until late 1953.[16]

Most shows centered on Marshall's interviews with guest experts on topics like atomic structure, spectroscopy, or astrophysics. One contemporary critic praised the series as an "entertaining exploration of seemingly simple facts of universal science which direct our daily actions, of scientific inventions which we use constantly."[17] Marshall experimented with various techniques to engage the viewers, attempting to show the moon through a telescope or using a microscope to observe "the dust on a butterfly's wing, and the hairy edge of a fly's wing."[18] Perhaps most important, he was sufficiently telegenic to appear also as a guest on other variety shows and to make commercials for the Ford Motor Company.

"Television is show business, not a lecture podium that can be forcibly poured by electronics into the living room," Marshall explained in an article in *Physics Today*.[19] Science, he acknowledged, must compete with other content; "a simple twist of the channel selector switch can replace a science presentation with a musical selection, a wrestling match, or a snappy news reel." To prove to viewers that "science is fun," he wrote, "I must wheedle, cajole, persuade.... Something—in my own manner or their genuine interest in knowing what this science stuff is all about, anyway—must prevent their turning to another channel."

In 1948 Baltimore's newest television station, WMAR-TV, was also eager to fill its broadcasting hours, and journalist H. L. Mencken suggested that the managers ask Watson Davis for ideas "for live shows on scientific subjects."[20] The station was, in fact, already discussing a series with the director of public relations at Johns Hopkins University. After graduating in 1937 with an M.A. from Western Reserve University, Lynn D. Poole had handled publicity for the Cleveland Museum of Art and the Walters Art Gallery in Baltimore, and served as a military public relations officer during World War II.[21] When he was mustered out in fall 1945, he returned to Baltimore and was hired by the university's provost to help place the faculty in front of the public and to promote the institution's name and reputation.[22] As director of the new public relations and news office, Poole created a television series that opened with a panoramic view of the campus and projected an image of science as part of culture, intellectuals as trustworthy authorities, and Johns Hopkins University as the hub of academic, scientific, and cultural expertise.

Unlike previous broadcasting popularizers, Poole was neither scientist nor science journalist nor educator, and thus represented a new phenotype. He had not initially intended to host the show himself but when a guest balked and insisted that Poole join him on camera, the slender, urbane man proved to be appealing to viewers and so became the host as well as producer and principal writer.[23] On screen, Poole added a cool, reserved presence, exuding academic authority even when he misspoke, always seeming to be part of the university family without claiming special expertise. He was sufficiently at ease even on a live show to crack his knuckles and smoke cigarettes. He never adopted the role of mere "announcer," who nods his head and deferentially introduces each expert. Instead, Poole created the persona of the scientist's facile promoter and authoritative interpreter. He continued that role in later years as author of dozens of popular books such as *Today's Science and You* (1952), *Frontiers of Science* (1958), and *Ballooning in the Space Age* (1958).

The Johns Hopkins Science Review premiered on March 9, 1948, and later that year became the first university program, on any subject, to be shown weekly on network television. In 1950 the series moved from the CBS affiliate to Baltimore station WAAM-TV and was broadcast via the DuMont network on weekday evenings until 1954, and then on Sunday afternoons until March 1955, eventually winning two prestigious Peabody Awards. Poole also produced three other series (*Tomorrow*, March–June 1955; *Tomorrow's Careers*, 1955–1956; and *Johns Hopkins File 7*, 1956–1960), which included some attention to science.

FIGURE 21. Actors and television host Lynn Poole (far right) on set of *The Johns Hopkins Science Review*. This September 1953 episode, "3-D in Science," described entertainment and scientific applications of stereo-vision. Poole is demonstrating a device invented by Oliver Wendell Holmes. Courtesy of Ferdinand Hamburger Archives, The Johns Hopkins University.

Every week, in over three hundred programs, *The Johns Hopkins Science Review* promised viewers a "look over the shoulders of today's scientists" to "catch a glimpse of the results of their work."[24] The topics ranged from archeology to public health to zoology. Shows like "All about the Atom," "The World Is an Atom," "What Is an Isotope?" (a 1952 episode on which a physicist drank iodine-131 to demonstrate how to measure radioactivity), and "The Peaceful Atom" documented the atomic age. Poole and his guests asked why muscles got tired, what causes cancer, why adding fluorine to drinking water prevents tooth decay, and what could be learned from solar eclipses; they demonstrated oscilloscopes and electron microscopes. Usually the guest scientist would explain how innovative equipment operated, and Poole sometimes conducted small experiments himself or engaged in stunts (such as being spun in a chair to demonstrate centrifugal force).

OLD MESSAGES, NEW (AND OLD) IMAGES

Lynn Poole wanted to create programs that "would entertain, delight, and hold audience attention while giving out worthwhile information," but he was exceedingly aware that "if informational programs are to survive, they must be planned and presented in such a way that they can hold their place in competition with the mystery drama, variety show, and quiz program."[25] Acknowledging this entertainment focus did not, Poole admitted, necessitate "having '50-Beautiful Girls-50' representing atoms and molecules on a show dealing with chemistry" or "making a monkey out of a famous scientist or turning his presentation into a circus."[26] Instead, he and his staff attempted to exploit more dignified techniques to reach their viewers, using familiar household objects, historical anecdotes, photographs, and charts.

Two award-winning episodes from the 1954–1955 season exemplify the diversity of approaches. In "Toys and Science," a toy maker demonstrated the scientific principles behind various toys, from a simple spinning top to kaleidoscopes and mechanical helicopters: "It isn't often that we stop to think that toys . . . embody principles discovered by Archimedes and Newton."[27] "Conquest of Pain" used still photographs to illustrate the history of anesthesiology: "Montgolfier had invented a balloon to carry above the earth" [photograph of a balloon], plans were under way for the first steamboats [photograph of steamboats], homes were lit with gas lights [photograph of gas lights], and "science was moving forward."[28] Later in the program, a physician demonstrated the administration of modern anesthetics.

No texts existed to guide these entrepreneurs, so Poole published *Science Via Television* (1950) to document their innovative presentation techniques. To televise living microorganisms in a drop of polluted water, the Hopkins crew projected a microscope image onto a translucent paper screen and then focused the camera on the image from the opposite side. In other programs, the camera lens was simply placed on top of the microscope. Live broadcasts were unforgiving, however. Even writing on a blackboard had to be rehearsed carefully lest a scientist move too quickly for the camera to track. Programs could not run a second too long—or too short. Poole also had to worry about the appropriateness of illustrations, especially for medical topics that audience members might find offensive. Just enough detail. Not too much blood. "Viewed scientifically," Poole wrote, films or photos of surgery might be

FIGURE 22. Demonstration of group therapy techniques staged for *The Johns Hopkins Science Review*, December 1951. The episode, called "Troubled People Meet," featured an appearance by Johns Hopkins University psychiatry professor Jerome Frank. The series producer and host Lynn Poole is visible at left, standing off-camera. Courtesy of Ferdinand Hamburger Archives, The Johns Hopkins University.

"exciting" and "viewed artistically they have aesthetic merit," but a layperson unused to such images might find them "revolting."[29]

Money posed the most formidable challenge. Without institutional commitment and resources, or commercial or network sponsorship, even a well-produced and award-winning educational program could not continue indefinitely. For *The Johns Hopkins Science Review*, public relations goals proved to be key, just as they had been for the Smithsonian radio programs. For the first year and a half, the university underwrote all production costs. To reduce expenses, Poole became skilled at enlisting volunteers. Then the university slowly began to accept "grants" and donations from both the network and the local station. The DuMont network provided annual donations that increased from $10,000 to over $50,000; the local station made other cash donations and, in 1952–1953, endowed the WAAM Television Fellowship at Hopkins.[30] For eight months, the university also accepted commercial sponsorship from a chemical company, although other potential sponsors were rejected:

> Many companies have made offers, but the University must be selective. It doesn't want "middle" commercials, and the advertising must be institutional with no product being offered for sale.[31]

Even as he resisted paid advertising, Poole accepted cooperation, donated equipment, and assistance from government agencies and companies like General Electric, General Motors, Monsanto, and RCA, frequently mentioning the names of product manufacturers on the air. Credit was "always given to the industry," Poole's longtime producer explained, and "industry has always been willing to cooperate under these terms."[32] In 1955 direct expenditures were said to be around $30,000 for fifty-two programs but this amount did not include the salaries of Poole and his office staff, nor did it account for the donated time of scientists, technicians, and volunteer actors. Publicity emphasized the theme of "science on a shoestring," comparing the small amounts spent for the Hopkins series to the huge costs for network shows and emphasizing that the program earned "no revenue for anybody" and was like "an awkward country girl competing with the big city's smoothest."[33] This carefully constructed image of amateurism was designed to resonate with Americans' reverence for stories about humble origins and trust of the rural Mr. Smith when he comes to Washington. Science, this television series declared, was all about knowledge and never about money.

As with earlier radio programs, television also provided powerful messages about who could (or should) be a scientist. In the 1930s, radio shows only occasionally featured the work of female scientists; women were, after all, still a minority of the overall scientific workforce in the United States. Watson Davis did interview a few women through the years on *Adventures in Science*, and the annual radio programs featuring winners of the Science Talent Search always included at least one (exceedingly bright) young woman. For its time, *The Johns Hopkins Science Review* conveyed a more discouraging (and false) message, however. The guest lists of the four series that Poole produced from 1948 to 1960 show that women were far more likely to appear as models conducting demonstrations (sometimes costumed in bathing suits) than as guest scientists. Only about 2 percent of all the guest experts were women scientists or physicians and of these women, only two (Cornelia Snell and Charlotte Silverman) were a program's featured guest.[34] Hundreds of male scientists shared the stage only with Poole, but eminent researchers

like parasitologist Eloise Cram, industrial hygienist Ann Baetjer, solar energy pioneer Maria Telkes, and archeologist Betty J. Meggers appeared as
part of groups, sharing attention and camera time with male colleagues or
technicians.[35] Women, the series consistently declared, were not science's
stars, only part of the team. When Hopkins professor Isabelle Schaub (author of a major textbook in diagnostic microbiology) was Poole's guest, she
discussed how young women could train to be technicians, not how they
might aspire to be scientists or professors. This aspect of television science—
the absence or marginalization of women scientists as the face as well as the
voice of science—continued for decades.[36]

THE MARKET FOR EDUCATION

The economic environment for education-oriented television during the 1950s
provides another answer to why popularization eventually leaned more
toward "show business" and away from depicting the intellectual business
of science. By partnering with a local station and then with a national network, and by accepting donations from each, Johns Hopkins University had
parlayed its investment and reached a national audience without having
to build and operate its own television facilities. Few other universities or
scientific associations could afford to do likewise.

Although many U.S. universities established nonprofit television stations,
usually in support of mass communications programs, significant commit-
ments from trustees (or legislatures) and additional foundation grants were
necessary to support even modest studios for regional broadcasts. In 1952
the regents of the University of Wisconsin were informed that a reasonably
well-equipped studio laboratory would cost at least $100,000 and a state-
wide educational system, over $3 million.[37] Because university operations
competed for local viewers, commercial stations even attempted to discourage the federal government from reserving any channels for educational use.
The Federal Communications Commission (FCC) informed University of Wisconsin officials that they should accept their educational television channel
allocation quickly in 1952 because commercial interests were applying pressure to acquire unused channels.[38] When the FCC decided to allow nonprofit
groups to acquire television licenses, no provision was made for federal funding of production facilities or for transmission towers as powerful as those
of the commercial stations; federal subsidy was made available only around
1975 to supplement state and private funding. By 1954 estimates of the cost of
starting an educational station, depending on capacity and equipment, ran

from \$33,000 to \$750,000, with annual operating costs between \$25,000 and \$500,000.[39]

Commercial network programming was also shaping viewing habits and audience expectations for what television should be. Universities had to consider, one observer wrote at the time, whether the investments were justifiable if educational television could not "please a viewing audience that seems to be perfectly content with the offerings of wrestling, westerns and deep décolletages."[40]

Some scientific groups attempted to produce science-focused educational programming and, as happened with radio in the 1920s, newly established television stations sought out local talent. Regional sections of the American Chemical Society assisted in developing "Science Page" segments for the *Magazine of the Week* series on KTLA-TV in Los Angeles; WMAL-TV in Washington, D.C., created a series around scientists at the National Bureau of Standards; and WMAR-TV in Baltimore produced in summer 1949 a physics series called *Atomic Report*.[41] Local educational programs, however, usually lacked sophisticated sets or urbane interviewers and offered weak competition for clowns and cowboys.

When the California Academy of Sciences ventured into television in 1949, its series *Science in Action* was hosted by Earl S. Herald, curator of aquatic biology. Underwriting from a local bank allowed the academy to produce elaborate programs on a wide range of topics—"from the inevitable atoms, anesthesia, bison, and calligraphy all the way to wombats, Yampa canyon, and zoo-morphology."[42] Early series guests included astronomer Harlow Shapley and a half dozen Nobel laureates. These scientists were not asked to lecture; instead, as with *The Johns Hopkins Science Review*, the visitors performed such tasks as isolating viruses, grading eggs with a micrometer, or assembling a transistor or else explained their research to Herald in conversational tones.[43] Sets were intentionally constructed to be "intimate, unostentatious and modest in feeling, in keeping with the spirit of scientific pursuits."[44]

Programs like these took advantage of the fact that television was usually viewed at home on a small screen. "The limitations of live television, enforced to a degree also by the small size of the screen on which it is viewed, are no particular obstacle to our specific aim—an informal, close-at-hand atmosphere where the viewer sees scientists at work in their natural surroundings," academy executive Benjamin Draper explained.[45] The surroundings were not really natural, of course. The laboratory in which Herald and his guests stood—which audiences undoubtedly presumed to be at the California Academy of Sciences—had been constructed inside a television studio.

Such artistic deception was deemed acceptable. The scriptwriter for a television science program, Poole explained, must always strive for a sense of reality, even while he is engaged in creating an illusion of actuality and developing "an atmosphere of here and now, an atmosphere leading the viewer to the feeling that he himself is beside the demonstrator."

> One of the best ways to do this is to make no mention of the "area" in which the program is being given. If the program is confined to an uncluttered, small space in which the action takes place at rapid pace, the viewer will not think about the location of the action.[46]

Exploiting a little drama, adding a few celebrities, some illusion, and an occasional boa constrictor helped to retain viewers' attention. Three years after its premiere, *Science in Action* continued to draw praise as an educational series and was competing successfully for evening audiences, but it remained a regional production.

In 1953 the American Museum of Natural History in New York launched a more ambitious, though short-lived, national series, *Adventure*, in cooperation with CBS. The network handled all aspects of production and gradually seized more control over content. During the first season, *Adventure* emphasized the research and exhibitions of the museum's curatorial staff. The premiere on a Sunday afternoon featured "a trip through space via the Hayden Planetarium," "Life in the Garden" (a film about competition among species), and "Undersea Story" (a film exploring the lives of Sicilian tuna fishermen). Segments of subsequent programs looked at bees, the Saharan desert, chicken embryos, human sensory perception, and Mayan gold artifacts, but within a few months, programs had begun to focus on a single theme and include more films produced especially for the series (e.g., "Army Ants" and "Celestial Hide and Seek," about eclipses). There were interviews with well-known scientists like Alexander Fleming and Konrad Lorenz, episodes that re-created the discovery of cortisone and Darwin's voyage on the *Beagle*, and travel films about Alaska, New Guinea, and Tibet. *Adventure* showed an early Jacques-Yves Cousteau film about the mating habits of fish; another episode ("Catalina Under Sea") was reportedly the "first live undersea broadcast in television history."[47] Topics like these, historian Gregg Mitman points out, had a social purpose aside from entertainment; they encouraged "audiences to draw universal moral principles from nature" and reaffirmed "the values of family life," although not every show centered on sentimental representations of flora and fauna.[48] Science was also placed in its political context. In 1955 *Adventure* looked at science's atomic legacy—visiting the testing facility at Yucca Flats, Nevada, and discussing the long- and short-term effects of radiation.

Adventure's final season offered a smorgasbord in which science and culture were displayed side by side. Historian and novelist Bernard de Voto toured the American West and described the cliff dwellings at Mesa Verde. Comedian Henry Morgan hosted programs about human physiology. There were programs on seals, penguins, apes, headhunters, rattlesnakes, hawks, and horses, and four episodes on the world's great religions. The last show in 1956 further blended fiction and fact by including film clips from a new movie (*Moby Dick*) to illustrate discussion of whales and whaling, thereby foreshadowing movie tie-ins now commonplace throughout television.

Even though they attempted to cater to television's entertainment trends and were often praised by the broadcast industry, programs like these were gradually shoved out of prime-time hours by comedies, game shows, and action dramas that drew larger audiences and earned more advertiser dollars. The first major content analysis of programming, sponsored by the Ford Foundation in 1949–1951, showed that advertising was consuming 20 percent of television time and educational programs less than 1 percent, a situation that worsened throughout the 1950s as network-produced shows began to dominate prime-time hours and further limited viewers' choices.[49]

COMPETING WITH CIRCUSES AND WIZARDS

"This is no time for modest aspirations," political scientist Harold D. Lasswell admonished educational broadcasters in the early 1950s.[50] Nevertheless, most educational programming for adults continued to favor traditional lecture or discussion formats, such as television "college courses" offered for credit. *The University of the Air*, for example, began in 1951 as a cooperative venture involving Triangle Television and twenty-two Pennsylvania academic institutions, and then expanded nationwide with presentations like Thomas P. Merritt's "Nuclear Physics" (1951) and Russell C. Erb's chemistry course, taught every semester during the 1950s.[51] Harvey E. White's physics course on NBC's *Continental Classroom* was reportedly the first course to be offered nationwide for college credit. Two months into the series, about five thousand students had enrolled and there were over a quarter million regular viewers. CBS's *Sunrise Semester* also featured courses by many prominent scientists.

When game shows like *What's My Line?* and *I've Got a Secret* became wildly popular, a few science-related shows attempted to piggyback on the fad.

What's On Your Mind? (ABC, 1951–1952), a prime-time panel show quickly retitled *How Did They Get That Way?*, was a public service program disguised as entertainment, addressing common psychological problems like gossip, hostility, and rejection. *What in the World?* (CBS, 1951–1955; broadcast locally until 1965) adopted a loftier tone. Host Froelich Rainey, director of the University of Pennsylvania Museum, challenged his academic guests to identify the origin, intended use, and circumstances of discovery for various archeological and cultural artifacts.[52] Philip Hamburger, the *New Yorker*'s television critic, marveled at "the knowledge these men display" and the "eloquence and facility with which they display it."

> They are relaxed and at ease as they turn [an item] over in their hands, put on and take off their glasses, cogitate a moment, and then come out with some of the damndest pieces of information a layman has ever seen. The gulf between this panel of experts and the average television panel is wider than the ocean itself.[53]

Science programs directed at children followed a different path, primarily because they served television's economic interests. On the assumption that such shows would encourage families to purchase television sets, networks initially gave children's programming high priority among the sustaining programs.[54] As advertisers recognized children as a potential market, this programming became quite lucrative for the networks. In 1949, 42 percent of children's programming was sustaining (without commercial advertising); during the next ten years, the proportion with advertising steadily increased.[55] ABC was the first network, in August 1950, to begin Saturday morning programming for kids, with shows like *Animal Clinic* and *Acrobat Ranch*. Within a year, the four networks together were showing twenty-seven hours of children's programs each week.[56]

Only a few of the early children's series focused wholly or in part on science, and those tended to emphasize fantasy and magic rather than facts and research. On *Science Circus*, during summer 1949, Bob Brown played an "absent-minded science professor" who performed various tricks for a live studio audience. *Mr. I Magination* (CBS, 1949–1952) included discussions about inventions whenever its host, Paul Tripp, took viewers, via innovative camera techniques, to "Ambitionville." *Kaleidoscope* (NBC, 1948–1952) consisted of narrations of instructional films on science, nature, and technology by *New York Times* sports columnist John Kiernan, who had become a national celebrity as a wisecracking "resident genius" on radio's popular *Information Please.*[57]

None of these series, however, matched the impact of *Watch Mr. Wizard*, whose winning formula—a "scientist" who demonstrates simple experiments to a youthful assistant—attracted loyal audiences for decades. Each week from 1951 until the 1990s, Don Herbert, in a succession of television projects, played the kindly "Mr. Wizard," explaining one or two scientific principles using experiments that rarely required complicated equipment.[58] The first assistant, ten-year-old "Willy," was one of Herbert's neighbors, but he was soon replaced by a child actor who could reliably memorize his lines. By 1953 girls as well as boys were assisting Herbert, each exclaiming "Gee, Mr. Wizard!" at the appropriate moment. At its peak, *Watch Mr. Wizard* drew about eight hundred thousand viewers per episode, but it had an even wider impact. By 1956 over five thousand "Mr. Wizard Science Clubs" had been established, with total membership over a hundred thousand. Teachers incorporated program themes into their classes, and "Mr. Wizard" science kits, books, and other product tie-ins filled the holiday gift lists of countless children.

Herbert epitomized the new hybrid science host. He had initially planned to be a science teacher and majored during college in science, English, and dramatics. After serving in World War II, he capitalized on his dramatic training and became a radio actor, first developing the Mr. Wizard character in the late 1940s around his own interests and abilities and then looking for a market. He heard that the Cereal Institute, a trade organization of breakfast food manufacturers, wanted to sponsor an educational television show: "They paid for the *Mr. Wizard* facilities... and NBC provided the time."[59] Throughout his career, Herbert maintained cordial relationships with commercial sponsors and underwriters, even appearing (in the character of Mr. Wizard) as the "General Electric Progress Reporter," who explained common household technologies, during commercial breaks of the *General Electric Theater* drama series. General Electric and its advertising firm liked Herbert because he had—in the company's own words—"a businesslike mien and an ingratiating air."[60]

A few science programs for adults during the early 1950s rejected such entertaining wizardry. On ABC's *Horizons* (a live Sunday evening broadcast, 1951-1955), Columbia University professor Erik Barnouw conversed with guests like Margaret Mead (who discussed "The Future of the Family"). During its last season, that series was refocused and retitled *Medical Horizons*. On CBS, another Sunday evening series, *The Search*, showcased work at U.S. universities, such as research on stuttering (University of Iowa), meteorology (University of Chicago), child development (Yale University), noise abatement (UCLA), race relations (Fisk University), and robots (MIT). The epitome of that decade's erudite television was *Omnibus*, a banquet of "information, enlightenment and education" for all ages, underwritten by the Ford Foundation, which consistently entwined science with culture.[61] For seven seasons, musical and dance performances, and dramatic scenes from *Antigone* or *The Mikado*, alternated with documentary segments about science, such as the demonstration of a diagnostic X-ray machine at the University of Rochester Medical School or an "atom smasher" at Columbia University. From 1948 through the late 1950s, serious programs like these celebrated scientists' accomplishments and projected a message that science was integral to society. Television had, however, begun its transformation into an industry preoccupied first and foremost with delivering profitable entertainment, not promoting civic education.

DISNEYFICATION

In 1954 a series premiered which transfigured the context for science popularization for decades to come. In a decision calculated to finance as well as publicize his new theme park, Walt Disney created a television series that was fast moving, beautifully animated and filmed, and family-friendly, quite unlike most programming then available.[62] Science (primarily science-in-nature) and technology (usually technological innovation) were regular themes from the outset, starting with the third broadcast, which featured excerpts from the studio's feature films *Seal Island* (1949) and *The Vanishing Prairie* (1954), with "behind-the-scenes" narration describing how Disney cinematographers captured time-lapse sequences of flowers blooming or insects in flight. Programs often combined live-action drama and animation into what Disney called "science factuals" (in contrast to science fiction).[63] Segments about real animals were paired with excerpts from cartoon fantasies like *Pinocchio*; discussions of space travel followed talking mice and singing cowboys. During the first twenty-one seasons, about one-fifth of *Disneyland* programs dealt with science, nature, space, or technology topics, the majority of those in "Adventureland" (nature) or "Tomorrowland" (space) segments.[64]

Tomorrowland segments were a mélange of "information and humor, pedagogy and entertainment" intended to celebrate the inventiveness of scientists and engineers.[65] Productions often involved notable experts. Rocket scientist Wernher von Braun and science writer Willy Ley, for example, helped to develop "Man in Space," "Man and the Moon," and "Tomorrow the Moon" programs, which combined newsreel footage of the U.S. space program with animation and mock-ups of imaginary space installations.[66] Another Tomorrowland project, "Our Friend the Atom," sought advice from Glenn Seaborg, Edward Teller, and E. O. Lawrence.[67] Chagrined at what he regarded as negative public attitudes toward atomic energy, Disney had wanted the project to demonstrate "limitless peaceful uses of atomic energy," in contrast to the "frightful terror ... sinister threat, mystery and secrecy" of the "military atom."[68] Using live-action footage and state-of-the-art animation, "Our Friend the Atom" depicted atomic energy as a colorful and powerful genie who could not be forced back into the bottle but could "be put to use for creation, for the benefit of all mankind."[69] The tone was almost wistful: Now that humans possess this terrible knowledge, they should be resigned to its reality and must direct the power toward peaceful uses.

When the Bell Telephone System decided to underwrite television science specials directed at family audiences, *Disneyland* thus provided a guide to success.[70] Bell and its parent company, AT&T, had been sponsoring cultural broadcasts for many years. Science was a natural choice for the new project, given the accomplishments of the company's research group, Bell Laboratories.[71]

To make the series, the company's advertising firm chose a leading Hollywood director, Frank Capra, famous for *Mr. Smith Goes to Washington* (1939) and *It's a Wonderful Life* (1946). Capra directed the first four broadcasts in the series (*Our Mr. Sun*, 1956; *Hemo the Magnificent*, 1957; *The Strange Case of the Cosmic Rays*, 1957; and *The Unchained Goddess*, 1958). The next four (*Gateways to the Mind: The Story of the Human Senses*, 1958; *The Alphabet Conspiracy*, 1959; *The Thread of Life*, 1960; and *About Time*, 1962) were created by Warner Brothers Studios using similar themes and styles.[72] In this self-described project of "public education through entertainment," the goal was to convey the importance of basic science and the nobility of scientists' quest for knowledge.[73] Each program combined clever animation with careful technical explanations from real scientists.

The productions have had unusual influence because of their longevity in the cultural marketplace and their multiplicity of use. Premier and repeat television broadcasts were followed by over forty years of continued replay in public and private settings. An educational outreach campaign underwritten by Bell Laboratories and AT&T provided free or low-cost copies to schools. Over sixteen hundred prints were in circulation during the late 1950s and early 1960s, viewed within that first decade by almost five million elementary and secondary schoolchildren and over half a million college students.[74]

Film and video copies remain in countless school libraries, and sets are still available for sale through educational publishers.

As with the Smithsonian and Science Service radio broadcasts, the involvement of prestigious scientists lent authority to the Bell productions. An advisory board, cochaired by retired Bell Labs engineer Ralph Bown and Rockefeller Foundation mathematician Warren Weaver, included distinguished experts in acoustics, anthropology, microbiology, and physics, and these men provided criticism on all scripts. For each program, additional specialists were invited to review content, and other scientists appeared in filmed segments.

No one—neither directors, corporate underwriters, nor scientists—willingly relinquished control to any other party, creating a contested situation that historian James Gilbert has well documented.[75] Capra's original contract required him to "satisfy" the advisory board chairman and vice-chairman, as well as any special scientific consultants, yet granted him artistic control over the final cut. Disputes on the early scripts (and they were numerous, according to Gilbert) thus tended to be resolved in Capra's favor. The success of *Our Mr. Sun* also put Capra in a favorable position to negotiate more control over the second script, including the right "to ignore scientific advice should it compromise the entertainment value of the series."[76] Although scripts still had to be approved by AT&T and its advisory board, Capra then had the option of shelving a program if he believed any final changes would "destroy" its entertainment value.[77] When tension with the advisory board (and cost overruns) continued, as Gilbert shows, Bell's advertising agency intervened, asserted more control, and eventually shifted production to Warner Brothers.

FACES AND VOICES

Science's public face was changing. Scientist-hosts gave way to celebrity intermediaries. Although he did not receive top billing in the first special, Frank Condie Baxter, a balding, bespectacled literature professor from the University of Southern California, quickly became the centerpiece of the Bell productions. Baxter had recently hosted two award-winning educational television programs on Shakespeare, and now, as a character called "Dr. Research," he dominated the presentations, serving as both host and principal narrative voice.[78]

On camera, Baxter was alternately shy and sharp, approachably friendly, earnest, and innocuous, radiating concern that the audience understand the science. As had happened with Lynn Poole, many viewers probably assumed that Baxter was a scientist even though he was never described as one. Eventually, American popular culture adopted Dr. Research as science's symbolic mouthpiece, the screen personification of everyone's favorite science teacher.[79] In *Our Mr. Sun*, the character was a jolly, smiling optimist standing somewhat stiffly in a rumpled tweed jacket. By the third program, he projected a more relaxed and polished personality, better dressed and more in command of his material.

Baxter's costar in the first program was comedian Eddie Albert; then, for the next three programs, the wisecracking, anti-intellectual "Writer" character was played by Richard Carlson, fresh from his role as an FBI spy in television's *I Led Three Lives*. The Writer provided a skeptical counterpoint to Dr. Research's self-assured proclamations. Each proclaimed science's importance, but the Writer always wanted to ask one more question.

As radio science programs had done, the Bell programs used music and other sound effects to heighten the dramatic impact. A theremin (an electronic instrument commonly heard on the sound tracks of 1950s science fiction movies) oscillated whenever the narrator described something "mysterious" or stated that scientists' knowledge was incomplete. Familiar passages from symphonies and Wagner's "Ride of the Valkyries" added to the sense of excitement. Beethoven's "Ode to Joy" played in the background when Dr. Research concluded:

> What a glorious opportunity to add to man's history . . . to harness those nuclear fires for man's use! To accept the challenge of creation. . . . and to use the gifts God gave us to explore the grandest of all frontiers . . . the Universe! For the more we know of creation, the closer we get to the creator.

Although later Bell productions favored 1960s pop tunes, the music of Beethoven and other classical composers still echoed whenever characters discussed faith, mystery, or myth.

The quality of the animation also ensured large audiences. UPA Studios, Shamus Culhane Productions (Culhane had worked on the Disney features *Snow White* and *Pinocchio* and many other films), Chuck Jones (of Bugs Bunny and Roadrunner fame), and Warner Brothers all created characters who explained scientific concepts or served as comic foils for Dr. Research and the Writer. Puppeteers Bil and Cora Baird created special marionettes for *Cosmic Rays*. Voices for the fictional characters were supplied by such Hollywood stars as Mel Blanc, Hans Conreid, Lurene Tuttle, Franklin Pangborn, and Lionel Barrymore (who played "Father Time").

The cartoons often reinforced negative stereotypes about women, as did the choice of actors and guest experts. All the actors were male, with the exception of a young girl who played "Alice in Wonderland," and few female scientists appeared on camera. Even the advisory board complained about the sexism in the cartoon character "Meteora" in *The Unchained Goddess*, but Capra refused to change the script, saying that Meteora's petulant personality was an appropriate characterization of weather's capriciousness.[80]

The success of the Bell specials influenced how television would present science from then on. Old-style interview programs or demonstrations now seemed too dull. Science could indeed be made exciting, entertaining, and enjoyable to children *and* their parents. The premiere of *Our Mr. Sun* attracted about twenty-four million viewers in Canada and the United States, almost one-third of the television audience (maximum viewing audience for *The Johns Hopkins Science Review* was then around half a million). Rave critical reviews appeared in the press, and the first show won an Emmy award. *Hemo, Goddess,* and subsequent programs in the series did comparably well.

CULTURE, VALUES, AND COLD WAR POLITICS

The same political, social, and religious issues that reshaped postwar research agendas affected the content of popular science on radio and television. Controversy could no longer be avoided altogether (as Austin Clark had advocated); science was caught up in acrimonious political debates, and the conduct of science itself was increasingly open to public questioning. Science popularization began to exploit an interrogatory approach, whether the topic was nuclear physics, climate change, or human biology.

In *Adventures in Science*, the juxtaposition of questions and answers had served to emphasize the differences between scientific and lay knowledge. In the Bell series, both questioner and answerer now constructed science as only one of many ways to understand the world. Inquiries about the natural world, the narrator explained, are limitless, while scientific knowledge itself is limited. Given enough time and resources, scientists will answer many of these questions and will raise even more, but scientists should never be presumed to know everything. Scientists, the programs implied, had become especially clever at identifying "what they don't know," while nonscientists tend to fill in the blanks with myths and superstitions.

In the most abstract and stylized of the Bell specials, *About Time,* Baxter interacts with two comic characters, the "King of Planet Q" and his aide, who are seeking advice on how to set their planet's clocks. "What *is* time?" Baxter asks. He then invites viewers to join him on "an imaginary trip into the vast world of stars, galaxies, meteors" to locate the answer. At a Planet Q observatory fitted with an "Earthscope screen," when a character wonders how Baxter knows that Einstein's theories about time and relativity are correct, Baxter suggests that they "ask a physicist—Dr. Richard Feynman of the California Institute of Technology." Feynman, standing in front of a blackboard covered in equations, delivers a complicated explanation. "But now I have more questions than ever," the king complains. "Good," Baxter responds smugly, repeating a series theme: "The more we learn, the more questions we have." "In the beginning, you had only one question," Baxter states, and that was "What time is it?" Now, thanks to science, humans can ask more sophisticated questions such as "How did time start?" or "Did the universe begin with a big bang?" or "Did the universe ever *have* a beginning?" Each discovery thus provokes more questions, more answers, more research, more confusion. In *The Strange Case of the Cosmic Rays*, Dr. Research explains that "as late as 1932, science had the universe all neatly wrapped up in three basic packages—electrons, protons, neutrons." In a mere twenty-five years, however, that number had "jumped to at least 20." No one knows how many more particles will be discovered, the Writer declares; scientists are drowning in knowledge ("a veritable deluge") and "crying for a Moses to lead them out of the wilderness of nuclear particles."

Such overt religious references percolate throughout the series, as do invocations of magic. Religion was used constantly to provide an alternate source of answers, sometimes with a certainty that science itself disclaimed. To an extent that would probably have astonished those scientists who debated antievolutionists in 1925, the Bell series promoted an image of a science that did not merely cohabit with religion but sometimes deferred to it. In the opening to *Our Mr. Sun*, a chorus sings a passage by Beethoven while a verse from the book of Psalms ("The heavens declare the glory of God") is superimposed on film of a glorious sunrise. The scene then shifts to a cluttered laboratory/office where the Writer asks, "Doctor, how was the sun born?" When Dr. Research replies, "We don't know exactly," the Writer exclaims, "You're a scientist, and you don't know?" Baxter (not confessing to his lack of scientific credentials) responds that guessing about an answer is for writers, not scientists. Then they demonstrate their differing approaches to seeking knowledge—on one side is Dr. Research's screen, reserved "just for facts," and on the other, the writer's "magic screen" (a manifestation of his imagination, where the cartoon characters always appear). "Your science and my magic," the Writer quips.

Animated characters and actors distinguish "facts" (the domain of science) from "mystery" (the domain of religion). The cartoon character "Father Time" states, "If you want facts, ask your scientist." Explaining that "primitive curiosity was the beginning of all your science," "Mr. Sun" bemoans the loss of myths and mystery ("Today, instead of temples, rituals, and hymns to the Sun, it's domes, gadgets, charts, and numbers. I'm demoted to a specimen"). Characters hint that myth might be intrinsically purer than science and that something important might be lost when science provides *all* the answers. In consolation, Baxter then reassures everyone that science will not destroy *all* of life's mystery, because "in seeking for knowledge" scientists too "are reaching for the great light beyond." At the conclusion of *Our Mr. Sun*, the camera focuses on a cross silhouetted against a sunset, choral voices swell, and the voice of old Father Time admonishes viewers to "go ahead, ask, inquire, seek the truth . . . measure the outside with mathematics but measure the inside with prayer." In *Hemo the Magnificent*, a cartoon character representing hemoglobin ("Hemo") refers to himself as "the sacred wine in the silver chalice." In *The Unchained Goddess*, Dr. Research declares that "of all his creatures, God has given only man the curiosity to ask why, the spirit to find out why, and the reasoning power to understand why," and later cites the book of Job to explain weather. The closing scene of *Gateways to the Mind* shows a white-robed chorus in the Hollywood Bowl standing in shape of a cross. The host explains, "How man chooses to use this knowledge in shaping his world will determine the future of mankind," and the chorus sings, "Amen, Amen." Even scenes without explicit religious symbols or Bible verses manage to convey respect for spirituality and religious authority. In *Hemo the Magnificent*, Hemo complains that humans cannot properly tell his story because they do not understand him: Blood represents life and should therefore not be overanalyzed or reduced to an experimental object.

Evolution was also laid open to question. In *Hemo the Magnificent*, the narrator explains that "water animals crawled out onto land" and certain cells specialized into lungs because "they had to learn how to breathe." "Now, wait a minute, Doc, wait a minute," the Writer yells. "Are you trying to say that I'm descended from some type of sea gnat?" Gilbert found that in the original script, Dr. Research was to have answered simply "No," but the scientific advisory board suggested "a stronger, more explicit response."[81] Capra compromised in the final version, having Dr. Research console the Writer and say: "You have a human spirit that separates you entirely from the animal world. But there's great mystery and great wonder in the fact that our body—this temple of the spirit—is built of billions of highly specialized cells."

Appeals to religious authority reflected the spirit of the times, of course. They also demonstrated a change in the acceptable parameters of popular science and how it was being shaped by its social context. Gilbert has suggested that Capra, a devout Catholic with a chemical engineering degree from California Institute of Technology, added these theological touches because he was convinced that improving public understanding of science would help ease cultural tensions between amateurism and expertise.[82] While Capra's strong views might explain the spiritual tone of the first specials, overt appeals to religion also appeared in the four created by Warner Brothers, and those scripts were approved by the same scientific advisory board. Only two of the eight programs did *not* contain Bible verses or similar religious references, and even those contained inspirational statements about "faith."

References to another dominant part of Cold War politics—secrecy regarding the development (and use) of atomic energy—also pervaded the films' dialogue and visual images. In *Our Mr. Sun*, a secretive animated character called "Chloro Phyll" posts signs on his laboratory door declaring, "Top Secret. Scientists Keep Out." "The Russians didn't invent me," Phyll insists. "I invented myself." In *The Strange Case of the Cosmic Rays*, an "Academy of Detection Arts and Sciences" committee is determining the "best detective story of the first half of the 20th century." The Writer enters "the strange case of the cosmic rays" into the contest, explaining that this scientific concept contains both mystery and adventure. When the committee—marionettes representing Edgar Allen Poe, Charles Dickens, and Fyodor Dostoyevsky ("Dosty")—responds that the winning story must have "scope," the Writer holds a photo of a mushroom cloud, and remarks, "Hold on to your hat. How's that for *scope?*" Later, Dosty dismisses science as "big doings about little nothings." The Writer counters that "out of little nothings like this have come most of the big doings of science," such as Newton and the apple (the law of gravitation) and "Maxwell tinkering with some tricky equations" ("a key to television") and the biggest "'something' man ever ran into—nuclear force."

Such offhand references to nuclear war infused all the Bell narratives. One cartoon character cheerily describes "nature's way to make a hydrogen bomb"; another is described as having the power of more than five million atom bombs; western bank robbers are dubbed "The Atom Bomb Gang," "The Uranium Gang," and "The Atom Bomb Boys." In *The Unchained Goddess*, the Writer explains that hurricanes release "more energy . . . every second than from a dozen atom bombs" and then (in apparent reference to contemporary proposals to use nuclear explosions to create artificial harbors) wonders what might happen "if we warm up Hudson Bay with atomic furnaces."

In the Bell programs, viewers were courted, attracted, and enticed to learn about science. With the best of intentions, in such broadcasts scientists and popularizers helped to construct an expectation of entertaining illusions. The images were no longer the uncontroversial, apolitical, and depersonalized ones projected by Clark and his contemporaries. Instead, the new mediated face of science attempted to resonate with the audience's revised ideas of science's negative and positive potential and with the reality of science's new politicized role in American culture. Attention to social, moral, and political aspects had become compulsory elements of popular science content.

MAGIC IN A BOX

Although the Bell programs influenced later presentation styles and tone, their success did not produce a flood of popular science series or specials on television. In 1954 the four networks broadcast nine regularly scheduled prime-time informational or educational science series. In 1958, after the demise of DuMont, the remaining three networks carried six series. In 1960, there were only two. Even the most creative projects had not been enough to lure viewers away from comedians and cowboys. Entertainment set the programming agenda.

Like radio in the 1920s, television was juggling two competing goals: its potential "as an inspirational medium" and the requirement that, as a business, it "reward its owners and advertisers."[83] Quality programming tended to survive only in time slots where competing programs lacked flash and glitz or when blessed with sufficient underwriting to be able to ignore ratings. Experienced popularizers like Watson Davis did not get involved in television, the scientific associations made only halfhearted attempts, and no latter-day Scripps with (very) deep pockets emerged to fund a new organization that would broker science popularization through television.

The economic structure of broadcasting also worked against attempts to encourage educational or informational programming. In the radio age, sponsors had shaped broadcast content by underwriting productions themselves or vetoing potential topics, pressure that eventually influenced the sustaining science programs.[84] When the networks began producing many of their own shows, advertiser influence was diluted but no less powerful. In the television era, the broadcasting industry became ever more cautious, imitative, and profit-driven, unwilling to risk losing a single night's audience. Rather than delivering content the public deserved or needed, the networks translated public whims directly into plotlines calculated to attract millions of viewers (and therefore advertiser sales). The lure of syndication, which could multiply the profitability of a series, further pushed content toward a "generalization" that guaranteed a program would be "acceptable to many different audiences in many different localities" around the country and eventually around the world.[85]

Marketplace pressure was nothing new, of course. Mass magazines and newspapers had long juggled advertiser demands and readers' changing interests with their own editorial standards and sense of public duty. What radio and television broadcasting brought to the marketplace was the ability to track and therefore react to audience preferences in (essentially) real time. During the 1930s, companies like Archibald Crossley's Cooperative Analysis of Broadcasting, C. E. Hooper, and A. C. Nielsen developed methods for surveying listeners via telephone and then synthesizing the data for use in planning the next season's (or next week's) programs. These tools helped to create a climate of perceived "egalitarianism" in broadcast content.[86] Radio seemed to be giving listeners what they wanted. Once Nielsen perfected a "black box" that could be attached to a radio (and later a television), broadcasters came to believe that they not only knew their audience's tastes but also could reliably predict their future choices. A herd mentality took over. Common wisdom in the industry accepted that popular science would not sell. By the 1950s, unless artificially disassociated from the advertising formula (either by subsidy or underwriting), science had little opportunity to compete in commercial network planning.

In seeing their audiences only "through a haze of statistical approximation," Thomas Streeter has pointed out, radio and television have favored flattened content and ignored subtle differences in viewers' personal interests, life experiences, or values.[87] To the broadcasting industry, the individual listener or viewer became, in effect, irrelevant.[88] Media consumers were interchangeable as long as they fit a particular demographic profile. Moreover, they were perceived as simultaneously buyers (investing time and attention whenever they "consumed" radio or television) and the thing sold (advertisers purchased time because they believed that consumers would watch the commercials cut into the entertainment or news). Networks created demographic profiles of who was likely to consume a program and sold time on the basis of that projection.

This context served to neutralize pressure for increased civic education. Edward R. Murrow, who had begun his CBS career supervising *Adventures in Science*, once predicted that television would "speak the truth as loudly as it will speak falsehood."[89] Murrow did his best to achieve that purpose, but *he* was the agent of change, *he* used the technology for that purpose. Neither his employer nor the industry was concerned with that mission. They were in the business of selling time.

Cultivating a scientifically literate populace never placed high on broadcasters' agendas, and from the 1920s on they thrust science to the sidelines. The old debate over who should bear responsibility for informing and educating the public about science could still be heard at educational and scientific conferences in the 1950s, while broadcasters continued to insist (understandably) that their "business" was not public education. At the end of the day, the broadcasters owned the control room and the microphone. Their illusions and actualities, their truths and falsehoods, their entertainments and commercial agendas, and their assumptions about what the audience wanted came to shape the popular science made available to all.

EPILOGUE

———— ✳ ————

Entertaining Lessons

It seems safe to say for any serious program there are at least one and a half million listeners in the United States who might try it—provided they do not have to give up a good laugh to do so.　　　　LYMAN BRYSON, 1945[1]

COMEDIES and cartoons of the 1940s and 1950s frequently depicted scientists as absentminded or "nutty" characters stumbling through their laboratories. By then, of course, most audience members knew how ludicrous such images were. Funny, yes; realistic, no. Scientists might occasionally be forgetful, unfashionable, or unsophisticated, but science itself was serious business, with significance for life, health, and survival and with consequences for all of humanity. The majority of popular science on radio and television had moved beyond a preoccupation with fact-based presentations and become more alert to the political and social ramifications of research. Remnants of the older format—descriptive talks delivered in tuxedos and pedantic accounts of "what we know" about echinoderms or cosmic rays—survived, but audiences and the commercial broadcasters had already expressed their preferences for documentaries laced with dramatizations and for the animated attention to atomic weapons typified by the Bell series.

In the mass media, science represented only one message among many, a statement of reason tucked amid music, laughter, sermons, sports, and soap operas. From coast to coast (and in a process duplicated around the world), radio and television had brought the voices of scientists and their interpreters to all races, genders, educations, incomes, and backgrounds, to any person inclined to listen. In the heat of August 1924, Washingtonians heard

239

about Magellanic Clouds; during the Great Depression, farmers in Missouri learned about the latest entomology research; on Halloween Eve 1938, a parody of science tricked listeners into believing that Martians had invaded New Jersey; and in the years following World War II, microphones trembled during atomic bomb tests and rattled with impassioned debate over who should control nuclear secrets.

Even with such variety, popular science occupied only a minor place in broadcasting schedules. The stations and networks initially welcomed scientists' contributions but then later refused to privilege them. If science programs could not conform to advertisers' preferences or to the average audience member's tastes, then they were relegated to the sidelines or eliminated altogether. In the United States, the federal government made little effort to alter that situation. The scientific community itself mustered only weak protest.

Now, in the twenty-first century, podcasting and similar new delivery technologies have overthrown the tyranny of scheduling and seem to promise increased opportunities for serious topics like science. Educational and informative programs need no longer be shuttled to the off-hours, nor be inhibited by censorship, nor be tied to commercial advertising, nor tiptoe around religious or political controversy, nor guarantee laughter or pathos. Producers can offer creative programming via the Internet at low or no cost; users can download talks, conversations, interviews, and dramatizations with liberty and then consume them at leisure. And yet, the outcome is no more assured that it was for radio, and the questions originally posed by E. W. Scripps and William E. Ritter remain relevant. What informational tools will people need to meet the challenges of the century ahead? How can popular science serve those needs? And, most important, who will produce, shape, and pay for popular science? The history of how science was popularized via radio and early television hints at likely outcomes, and also suggests some lessons.

CONTEXTUALIZATION

In the decades between Austin Clark's first radio broadcast in 1923 and his television appearance on *The Johns Hopkins Science Review* in 1952, the world outside the studio was reshaped by wars, political and economic crises, and cultural revolution. Great changes had occurred within science (and because of it). Science had become an indispensable, megabillion-dollar part of governments, industries, and universities. Legislatures no longer questioned whether research facilities were essential to their state universities. Taxpayers were underwriting generous funding for basic as well as applied research.

Over that same period, the physical devices that deliver radio broadcasts changed considerably, their cultural resonance expanding as their size and cost shrank. By the 1940s radios on shelves had displaced the large, expensive cabinets. The wooden box next to my grandfather's favorite chair dispensed Guy Lombardo's "Auld Lang Syne" at midnight every New Year's Eve. Later, a turquoise transistor radio brought Elvis under my pillow and a pink clock radio shattered each morning's dreams. Now, a podcast of a *Science Friday* radio show can be replayed during a Monday-morning subway ride. Radio has been transformed from novelty to furniture to personal accessory, and television has followed a similar pattern. The first postwar sets dominated suburban living rooms and social life; today, flat-panel screens seemingly float on the wall and "television" can be viewed on computers and cell phones.

What had appeared as an exotic toy in 1923 became a social necessity within merely a decade, and for each subsequent innovation the period of cultural assimilation has become shorter. Today we take radio and television for granted, much as the youthful Austin Clark did popular magazines. By the early twentieth century, the magazine—in which articles about Einstein and relativity were interspersed with celebrity interviews and short fiction—had become commonplace in homes and libraries. Magazines, however, were eighteenth-century creations made possible by ripening printing technologies, favorable economic conditions, and increased literacy; and the familiar format of mass-market magazines had emerged only in the nineteenth century.[2] That format had encouraged popularization of topics like science because a single issue could incorporate diverse divertissements—crisply written short fiction and lavish illustration followed by concise interpretations of specialist knowledge. Radio followed a similar (albeit accelerated) progression in which technological innovation combined with a favorable social, political, and economic context and offered up a mélange of content. Tuning in each week's *Adventures in Science* broadcast became like opening the latest issue of the *Saturday Evening Post*. Listeners and readers alike had certain expectations of quality and delighted in reliable novelty. Adding a little dramatic touch to a radio interview was like adding color to a magazine layout or an exclamation point to an article title. Audience acceptance grew, the business of producing radio matured, and the notion of listening to a broadcast became part of American culture. Every day, stations transmitted the aural equivalent of mass magazines, combining fact and fiction, politics and comedy, newscasts and Hollywood gossip, with orchestral accompaniment and applauding studio audiences.

IMAGINATIVE FAILURES

In 1925 Clark declared somewhat presumptively that "all of those who have talked over the radio . . . have acquired a wholly new concept of popular science."[3] To some extent, he was correct. The radio industry had challenged the assumptions that science must be addressed with dignity, that the best explanations took time and required cautious qualification, and that science should always be presented by scientists. The notion of *mass* communication of science—that is, of offering simplified explanations and condensed discussions to a broad-based, unselected, unfiltered, and unseen audience, and of doing so alongside entertainers and musicians (as if scientists were the third act on a vaudeville stage)—was indeed new. Unfortunately, many scientists appear to have perceived radio as more like a book or a formal lecture than, say, a popular magazine, where readers were free to choose or ignore particular articles, and might not read every word, yet would still renew the subscription. With expanded content choices, radio audiences gained control over experts who failed to understand their needs and preferences or failed to speak clearly. A simple twist of the dial could change the tune.

What seems astonishing in retrospect is how little the leaders of the scientific community seemed to respect broadcasting's power and potential. Although perceptions of popularization did change in the twentieth century, scientists' attitudes never achieved a stage of relaxed enthusiasm. The mainstream scientific community demonstrated little interest in grasping the opportunity to reach huge, diverse audiences, much less in acknowledging any professional responsibility to do so.

The owners and managers of the mass media ultimately controlled the switches, of course, but during radio's critical period, the scientific associations and leadership failed to lobby for more science programming. The scientific associations hung back, waited to be invited, cooperated primarily when they could control the content, and only grudgingly invested funds and efforts in popularization. With the exception of Science Service, most national scientific organizations either argued that radio should be reserved for educational purposes alone, held the microphones at arm's length, or attempted to exploit broadcasts primarily for fund-raising. A few universities and museums produced notable programs but these rarely had adequate funding and were usually confined to regional audiences.

Television sealed the deal. Preoccupation with entertainment, sensation-alistic performers, and a voyeuristic affinity for "actualities" did not seem a congenial environment in which to discuss the serious science-related issues of the day. Once television's ideal scientific personality had been culturally defined as Dr. Research, the task of convincing prominent scientists (whose cooperation remained essential for success) to use television for populariza-tion became all the more difficult. No one advocated flooding the airwaves with science. Instead, the broadcasts sanctioned by scientific groups were dispensed in discrete, carefully constructed packets, exhibiting scant faith in the audience's ability to choose wisely from a banquet of knowledge.

The scientific community also failed to capitalize on broadcasting's imag-inative potential, despite abundant evidence that these techniques could attract large audiences to serious topics. Orson Welles had successfully ex-ploited them in his programs; Edward R. Murrow painted elaborate word-portraits of the geographical and emotional landscape during the London blitz; famous playwrights turned their talents to scripting radio dramas; and in postwar radio documentaries, brilliant directors exploited entertain-ment techniques to encourage public debate on atomic energy. Programs sanctioned by the scientific associations tended to eschew these approaches or, in the case of *Adventures in Science*, to apply them tentatively. Too often, science was presented in unidimensional, highly intellectualized formats, drained of emotion and excitement, just as had always been traditional in professional print communication.

Economics played another role in the marginalization of science pro-gramming. Radio, Paul Lazarsfeld observed in 1942, represented a "conser-vative force in American life."[4] The audiences were self-selected. Broadcast-ing derived its income from advertising, which required networks to sustain the status quo, that is, to "avoid whatever deviates too sharply from what the listener already accepts" lest the dial be twisted elsewhere.[5] "Good" educa-tional broadcasts, Sydney Head has explained, invariably conflicted with the "basic principles of good commercial programming" by aiming at smaller audiences, costing more to produce, attracting little or no revenue, and erod-ing the ratings of programs scheduled before and afterward.[6] As a result, ra-dio content presumed to be even remotely educational was, as a matter of good business, relegated to channels, stations, or time slots that were neither main-stream nor prime time. Although, in theory, science could have been defined as part of broadcasting's public service obligation, even the entertaining sci-ence programs began to be consigned to off-hours because they discussed science. And as Watson Davis wryly observed, the greater public interest is not well served when science is "relegated to sunrise hours to be gulped like an awakening cup of coffee."[7]

With national networks firmly in command of the airwaves, science could only be included at the broadcasters' discretion. The federal government encouraged but did not require science, and the scientists themselves seemed determined to construct science programming as educational content. Laughter had no place in the public classroom. Each failed experiment—and scientists' continued resistance to using broadcast media—influenced the inclination to attempt another. When groups did step forward with sufficient money to hire the best creative talent and to produce accurate *and* entertaining programs, as did DuPont, U.S. Rubber, and AT&T, they did so out of motives linked to corporate public relations and at a level of investment the smaller associations could not match.

RESONANCE

Responding to the Scripps-Ritter challenge will not be easy in the twenty-first century. The audiences for popular science have changed, not least in the ways they consume such content and in the variety of sources delivering it. In the 1920s, families or friends clustered together around the wireless set; during the early days of television and of home access to the Internet, people practiced similar group reception. Today, information consumption is simultaneously solitary and social. Computer users may be sitting alone at the keyboard, but they are often engaging in social activity, interacting with a range of "potentially conflicting discourses" and sharing reactions with others.[8] Other media experiences are similarly social. Today's science popularizer, Milton Chen, has observed, must reject the assumption of "a single viewer selecting a science program to watch, sitting down on a couch in front of a set at the start of the show, and watching a complete program from start to finish."[9] Listeners, viewers, and users engage in multitasked consumption, connected to an array of devices. Each individual's ideas and images of science come not from a single text or broadcast but are continually plucked from the content of many.

The normal, fragmented, disjointed "background noises" of life, biologist and television host David Suzuki once pointed out, can easily distract us from intelligent selection and consumption ("from the moment we wake up to the time we go to bed, we are assaulted by information ... delivered in small chunks").[10] Science has not always risen above this noise—unless dramatized, fictionalized, personalized, politicized, or presented with humor.

With the addition of specialized cable, satellite, and Internet services to traditional radio and television, consumption and production patterns for people everywhere have grown more complicated. Understanding the audiences for popular science, and understanding how they assemble their beliefs about science, are essential first steps in understanding how to serve those audiences better. An updated, multilayered "encounter" model for popularization must assume multiple interconnections, multiple sources, and constant social reinforcement. Political scientist V. O. Key anticipated the complexity of communication in the twenty-first century when he wrote, "The messages of the media do not strike the isolated and atomistic individual; they strike, if they reach their target at all, an individual living in a network of personal relationships that affect his outlook towards the objects of the external world, including the mass media."[11] Just as researchers are developing more nuanced explanations of how the human brain works, why birds sing, and what creates earthquakes and tidal waves, those who would better understand the impact of popular science communication must acknowledge, per Tony Schwartz, that a message's cultural "resonance" (including alternative and contradictory information from other sources) may be as important as its particular content.[12] Radio's first audiences had to learn "how to listen." Today's wise consumers of popular science must be able to discern silly science from silly pseudoscience, sincere science from politicized science, uncertain but well-founded suppositions from deliberately deceptive half-truths. Teaching skills of discrimination suited to a multimedia, multichannel world may thus be as important as producing a few more well-crafted television shows or Web sites.

Participants in popularization today are also more difficult to characterize. Old models for popular science, based on assumptions of linear rather than simultaneous or parallel consumption of content, imagined sets of clearly identifiable communicators and receivers, speakers and audiences, gatekeepers and editors. For popular magazines, there were subscribers and readers, authors and editors, owners and advertisers. In Clark's time, the decision-making roles in broadcasting were relatively few. Clark chose the speakers and edited their talks. The local station manager's choices were constrained by economics and social convention. Listeners wrote letters directly to the guest speakers. And wearing a tuxedo could, indeed, seem to matter.

Today's multitudinous participants engage in intricate games and adopt many roles. Consumers of science news, information, and entertainment, especially via the Internet, confront multiple messages and multiple routes to information. The sources of what appears to be "scientific information" may have social, religious, or political goals that are unrelated to real science and that are deliberately obscured in the presentation. Those sources may seem as legitimate as the "California Astronomical Society" sounded to listeners of Orson Welles's Halloween Eve production. Navigating today's ocean of endless competing messages requires skill, skepticism, and the ability to resist the sirens, to resist the impulse simply to click, download, cross fingers, and hope for accuracy.

The dream advanced by E. W. Scripps and William E. Ritter remains a worthy one. Science's importance to daily and global existence has increased over the last century. In this context, moving steadily toward a goal of improved civic education in science should be obligatory, should be an effort in which scientists, government, broadcasters, and concerned citizens are all involved. The trick will be to remain flexible and anticipatory rather than stuck in antiquated patterns of communication that ignore changing audience needs and changing technological delivery systems.

In 1936 Frank Thone told members of the American Association for Adult Education that he believed people were "as eager as St. Paul's Athenians to hear some new thing" about science but that they preferred flexibility to pontification, including flexibility in how and *where* they chose to learn about science.

> Their Agora is the daily newspaper. It may be a less sociable institution than the Athenian market-place or the Victorian lecture-hall, but it is a much more flexible one. You can roll up a whole company of heralds, messengers, and gossips, stick them in your pocket, select the ones you want to listen to, and hear their stories whenever you please.[13]

What delight Thone would have had in selecting from the "heralds, messengers, and gossips" in today's media marketplace. I can just imagine the tall, red-haired botanist, striding through the streets of Washington, the latest portable communication device tucked in his shirt pocket, happily listening to the new voices of science.

Notes

Clark Papers (NMNH): Austin Hobart Clark Papers in the curatorial collections of Dr. David
Pawson, National Museum of Natural History, Smithsonian Institution (these records will
eventually be transferred to the Smithsonian Institution Archives).

Clark Papers (SIA): Austin Hobart Clark Papers, 1883-1954 and undated, Record Unit 7183,
Smithsonian Institution Archives.

Hale Papers: George Ellery Hale Papers, 1882-1937, Carnegie Institution of Washington (microfilm, Manuscript Collections, Library of Congress)

JHU News: Johns Hopkins University, Office of News and Information Services, Records, 1946-
(ongoing), Record Group Number 10.020, Special Collections, The Milton S. Eisenhower
Library, Johns Hopkins University

JHSR Videotapes: Videotapes of Johns Hopkins Television Programs 1948-1960, Special Collections, Milton S. Eisenhower Library, Johns Hopkins University

Merriam Papers: John C. Merriam Papers, MSS 32706, Manuscript Collections, Library of Congress

Millikan Collection: The Robert Andrews Millikan Collection at the California Institute of Technology (microfilm, Manuscript Collections, Library of Congress)

Poole Papers: Lynn Poole Papers, 1948-1976, MS 27, Special Collections, Milton S. Eisenhower
Library, Johns Hopkins University

Science Service Records: Science Service Records, Record Unit 7091, Smithsonian Institution
Archives

SI Press: Smithsonian Institution Press, Records, 1914-1965, Record Unit 83, Smithsonian Institution Archives

SI Secretary (RU45): Office of the Secretary, Records, 1890-1929, Record Unit 45, Smithsonian
Institution Archives

SI Secretary (RU46): Office of the Secretary, Records, 1925-1949, Record Unit 46, Smithsonian
Institution Archives

CHAPTER ONE

1. Austin H. Clark to Edward Wigglesworth, October 26, 1925, Clark Papers (SIA), 13:1.
2. SI Secretary (RU45), 46:7.

3. Charles D. Walcott to Austin H. Clark, October 9, 1923, SI Secretary (RU45), 11:17.

4. Clark spoke "with an unusually deep voice and a 'Harvard accent.'" "Austin Hobart Clark," *Lepidopterists' News* 9 (1955): 152. See also Clark's appearance on *The Johns Hopkins Science Review*, July 21, 1952, JHSR Videotapes.

5. Mark Sullivan, *Our Times: The United States, 1900–1925*, vol. 6 (New York: Charles Scribner's Sons, 1935), 422.

6. Ibid., 423.

7. Austin H. Clark to James McKeen Cattell, December 2, 1925, Clark Papers (SIA), 3:6.

8. Anthony Smith, *The Shadow in the Cave: The Broadcaster, His Audience, and the State* (Urbana: University of Illinois Press, 1973), 28.

9. Marshall D. Beuick, "The Limited Social Effect of Radio Broadcasting," *American Journal of Sociology* 32 (January 1927): 617.

10. U.S. Federal Communications Commission statistics, cited in Mitchell V. Charnley, *News by Radio* (New York: Macmillan, 1948), 6.

11. Susan Douglas, *Listening In: Radio and the American Imagination* (New York: Times Books, 1999), 52.

12. Watson Davis, "Government Control of Radio Telephony," *Scientific Monthly* 14 (April 1922): 397.

13. Douglas, *Listening In*, 52; Beuick, "The Limited Social Effect of Radio Broadcasting," 616.

14. R. W. Sorenson, "World Is Now in Radio Age," *Los Angeles Times*, April 2, 1922.

15. "Radio Owners' Census Is Begun in District," *Washington Post*, December 2, 1923.

16. "Analysis of Radio Broadcasting Audience Station WEAF," handwritten chart dated October 1923, Science Service Records, 89:2.

17. Douglas, *Listening In*, 64, 76. See also Michelle Hilmes, *Radio Voices: American Broadcasting, 1922–1952* (Minneapolis: University of Minnesota Press, 1997), xvii, 11.

18. Douglas, *Listening In*, 55–57.

19. Sorenson, "World Is Now in Radio Age."

20. Ellis L. Yochelson, *Smithsonian Institution Secretary, Charles Doolittle Walcott* (Kent, Ohio: Kent State University Press, 2001), 414–15.

21. Stanley Frost, "Radio—Our Next Great Step Forward," *Collier's Weekly*, April 8, 1922, 3–4, 18–24. See also Susan Smulyan, *Selling Radio: The Commercialization of American Broadcasting, 1920–1934* (Washington, D.C.: Smithsonian Institution Press, 1994), 32.

22. Charles D. Walcott to C. G. Abbot, December 20, 1918, SI Secretary (RU45), 35:13.

23. Notices for February 25 and 28, 1919, SI Secretary (RU45), 35:13. See also *Annual Report of the Board of Regents of the Smithsonian Institution, 1919* (Washington, D.C.: Smithsonian Institution, 1921), 12.

24. *Annual Report of the Board of Regents of the Smithsonian Institution, 1920* (Washington, D.C.: U.S. Government Printing Office, 1922), 26.

25. Walter Hough to W. DeC. Ravenel, February 17, 1920, SI Secretary (RU45), 56:15.

26. Charles D. Walcott to Austin H. Clark, October 9, 1923, SI Secretary (RU45), 11:17.

27. Russell Lynes, *The Lively Audience: A Social History of the Visual and Performing Arts in America, 1890–1950* (New York: Harper & Row, 1985), 39.

28. D. H. Killeffer, "Chemical Education Via Radio," *Journal of Chemical Education* 1 (March 1924): 47.

29. Afterward, Clark said that he had worked hard to devise "a tale about [the prince] and his activities for the newspapers so that they will not be tempted to overemphasize Monte Carlo and thus hurt his feelings." Austin H. Clark to Walter K. Fisher, March 22, 1921, Clark Papers (SIA), 5:7.

30. "Austin Hobart Clark," 152.

31. Interview of Lucille Mann conducted by Pamela Henson, June 9, 1977, Lucille Quarry Mann Papers, Record Unit 9513, box 1, Smithsonian Institution Archives.

32. See David L. Pawson and Doris J. Vance, "Austin Hobart Clark (1880–1954): His Echinoderm Research and Contacts with Colleagues," *12th International Echinoderm Conference, New Hampshire* (Lisse: Taylor & Francis, in press).

33. Austin H. Clark to Asa C. Chandler, November 20, 1923, Clark Papers (SIA), 3:7.

34. Austin H. Clark to Sir Sidney Harmer, January 11, 1924, Clark Papers (SIA), 6:5.

35. See David L. Pawson and Doris J. Vance, "On Board the *Albatross* in 1906: A Young Scientist Writes Home," personal communication, 2007.

36. Linda Elmore, "Finding Aid to Austin H. Clark Papers (RU 7183)," Smithsonian Institution Archives, n.d. See also Henson interview with Lucille Mann, p. 30. "Atom" quote: Austin H. Clark to C. C. A. Munro, May 20, 1928, Clark Papers (SIA), 8:8.

37. Ralph Edmunds to Austin H. Clark, October 22, 1923, Clark Papers (SIA), 5:3.

38. J. Walter Fewkes to Ralph Edmonds, October 26, 1923, SI Secretary (RU45), 46:7.

39. Copies of the first season's scripts contain handwritten comments like "Excellent!" from both Secretary Walcott and his deputy. SI Secretary (RU45), 8:46.

40. Annual Report of the Board of Regents of the Smithsonian Institution, 1925 (Washington, D.C.: U.S. Government Printing Office, 1926), 31.

41. Austin H. Clark to Ralph Edmunds, March 28, 1924, Clark Papers (SIA), 5:3.

42. Austin H. Clark to Charles D. Walcott, April 21, 1924, SI Secretary (RU45), 46:7.

43. Austin H. Clark to Charles D. Walcott, April 12, 1924, SI Secretary (RU45), 46:7. On Clark's letter, there are handwritten notations of approval from Secretary Walcott and Assistant Secretary C. G. Abbot.

44. Austin H. Clark to George N. Pindar, May 14, 1924, Clark Papers (SIA), 9:9.

45. Austin H. Clark to Frank Springer, June 9, 1924, Clark Papers (SIA), 11:6.

46. George N. Pindar to Austin H. Clark, May 12, 1924, Clark Papers (SIA), 9:9; Thornton W. Burgess to Austin H. Clark, June 10, 1925, and June 17, 1925, Clark Papers (SIA), 3:2.

47. C. C. Nutting to Austin H. Clark, November 14, 1925, Clark Papers (SIA), 9:6.

48. Many of these talks were reprinted in booklets; for example, Alexander Silverman, Charles G. King, Kendell S. Tesh, Alexander Lowy, Gebherd Stegeman, and Carl J. Engelder, *A Series of Seven Radio Talks on Chemistry and Human Progress*, Radio Publication 21 (Pittsburgh: University of Pittsburgh, 1926).

49. Killeffer, "Chemical Education Via Radio."

50. Smulyan, *Selling Radio*, 21–22. See also Judith C. Waller, *Radio: The Fifth Estate* (New York: Houghton Mifflin, 1946), 260–70; John C. Baker, *Farm Broadcasting: The First Sixty Years* (Ames: Iowa State University Press, 1981); and Jeanette Sayre, *An Analysis of the Radiobroadcasting Activities of Federal Agencies*, Studies in the Control of Radio no. 3 (Cambridge: Harvard University, 1941).

51. Frederic A. Leigh, "Educational and Cultural Programming," in *TV Genres: A Handbook and Reference Guide*, ed. Brian G. Rose (New York: Greenwood Press, 1984), 365; C. F. Marvin, "Radio Meteorological Services," *Annals of the American Academy of Political and Social Science* 142 (March 1929): 64–66.

52. W. W. Bauer and Thomas G. Hall, *Health Education of the Public: A Practical Manual of Technic* (Philadelphia: W. B. Saunders, 1937), 40. See also Smulyan, *Selling Radio*, 21–22.

53. Austin H. Clark to A. W. Greely, October 8, 1924, Clark Papers (SIA), 6:3.

54. Austin H. Clark to J. J. R. Macleod, December 19, 1924, Clark Papers (SIA), 8:4.

55. Austin H. Clark to Harlow Shapley, November 7, 1925, Clark Papers (SIA), 10:11.

56. Austin H. Clark to Edward Wigglesworth, October 26, 1925, and Edward Wigglesworth to Austin H. Clark, October 28, 1925, Clark Papers (SIA), 13:1.

57. Austin H. Clark to Thornton W. Burgess, January 11, 1926, Clark Papers (SIA), 3:3.

58. Austin H. Clark to Edward Wigglesworth, October 30, 1925, Clark Papers (SIA), 13:1.

59. Austin H. Clark to Thornton W. Burgess, June 20, 1925, Clark Papers (SIA), 3:2.

60. Austin H. Clark to Edward Wigglesworth, October 15, 1925, Clark Papers (SIA), 13:1.

61. Austin H. Clark to Thornton W. Burgess, June 20, 1925, Clark Papers (SIA), 3:2.

62. "Science Radio Talks" (typescript), May 31, 1924, Science Service Records, 74:4.

63. E. E. Slosson to George R. Mansfield, June 21, 1924, Science Service Records, 23:1.

64. E. E. Slosson to H. E. Howe, June 17, 1924, Science Service Records, 22:6.

65. Austin H. Clark to J. J. R. Macleod, December 4, 1924, and J. J. R. Macleod to Austin H. Clark, December 10, 1924, Clark Papers (SIA), 8:4.

66. Austin H. Clark to George N. Pindar, May 14, 1924, Clark Papers (SIA), 9:9; Austin H. Clark, "Radio Talks," *Scientific Monthly* 35 (October 1932): 354.

67. Austin H. Clark to Thornton W. Burgess, June 20, 1925, Clark Papers (SIA), 3:2.

68. Ibid.

69. Austin H. Clark to John C. Phillips, September 29, 1925, Clark Papers (SIA), 9:9.

70. Austin H. Clark to Thornton W. Burgess, June 1, 1925, Clark Papers (SIA), 3:2.

71. Clark expressed his views on race in various correspondence. See Clark Papers (SIA). He also was involved with (and a member of) the Eugenics Committee of the U.S.A.

72. Austin H. Clark to E. Lester Jones, December 1, 1924, Clark Papers (SIA), 7:6.

73. Austin H. Clark to Ralph Edmunds, January 15, 1925, Clark Papers (SIA), 5:3.

74. Austin H. Clark to Charles D. Walcott, April 3, 1918, Clark Papers (SIA), 12:5.

75. Austin H. Clark to James I. Hambleton, March 31, 1927, Clark Papers (SIA), 6:5.

76. Austin H. Clark to E. E. Slosson, May 19, 1924, Clark Papers (SIA), 10:11.

77. Austin H. Clark to James McKeen Cattell, December 2, 1925, Clark Papers (SIA), 3:6.

78. "Museums and the Radio," n.d., Clark Papers (SIA), 16:5. This is a draft of Clark's "Science and the Radio," *Scientific Monthly* 34 (March 1932): 268-72. See also 44-page carbon version in SI Secretary (RU45), 46:7. The folder is dated 1923, indicating when Clark began formulating these ideas.

79. Charles D. Walcott to Austin H. Clark, May 19, 1925, SI Secretary (RU46), 15:8.

80. Russel Nye, *The Unembarrassed Muse: The Popular Arts in America* (New York: Dial Press, 1970), 392.

81. Ibid., 392. See also Roland Marchand, *Advertising the American Dream: Making Way for Modernity, 1920-1940* (Berkeley: University of California Press, 1985), esp. 88-89.

CHAPTER TWO

1. Thornton W. Burgess to Austin H. Clark, October 1, 1926, Clark Papers (SIA), 3:3.

2. This summary of Burgess's early life relies primarily on his autobiography. Thornton W. Burgess, *Now I Remember: Autobiography of an Amateur Naturalist* (Boston: Little, Brown, 1960). See also Russell A. Lovell Jr., *The Cape Cod Story of Thornton W. Burgess* (Town of Sandwich, Massachusetts, 1974).

3. "Colonial Village at Springfield to Include Sanctuary for Birds," *Christian Science Monitor*, August 6, 1930, 5.

4. Thornton W. Burgess to Austin H. Clark, April 12, 1929, Clark Papers (NMNH). After one excursion, Burgess wrote, "We couldn't find but three [hens] and the thought that perhaps I was looking at the very last individuals of a species was rather sobering." Thornton W. Burgess to Austin H. Clark, April 30, 1928, Clark Papers (SIA), 3:3.

5. Burgess, *Now I Remember*, 141-42.

6. "WBZ Starts Radio Nature Association," *Christian Science Monitor*, February 18, 1925, 9.

7. Thornton W. Burgess to Austin H. Clark, May 27, 1925, Clark Papers (SIA), 3:2.

8. Burgess, *Now I Remember*, 145-46.

9. Ibid., 145.

10. Ibid.

11. Thornton W. Burgess to Austin H. Clark, August 11, 1925, Clark Papers (SIA), 3:2.

12. Austin H. Clark to Thornton W. Burgess, May 26, 1925, Clark Papers (SIA), 3:2.

13. Thornton W. Burgess to Austin H. Clark, May 27, 1925, Clark Papers (SIA), 3:2.

14. Austin H. Clark to Thornton W. Burgess, June 20, 1925, and Thornton W. Burgess to Austin H. Clark, June 22, 1925, Clark Papers (SIA), 3:2.

15. Thornton W. Burgess to Austin H. Clark, June 17, 1925, and July 10, 1925, Clark Papers (SIA), 3:2.

16. Thornton W. Burgess to Austin H. Clark, July 10, 1925, Clark Papers (SIA), 3:2.

17. Austin H. Clark to Thornton W. Burgess, June 1, 1925, Clark Papers (SIA), 3:2.

18. Thornton W. Burgess to Austin H. Clark, October 1, 1926, Clark Papers (SIA), 3:3.

19. Austin H. Clark to Thornton W. Burgess, June 16, 1925, and Thornton W. Burgess to Austin H. Clark, June 17, 1925, Clark Papers (SIA), 3:2.

20. Austin H. Clark to Thornton W. Burgess, June 16, 1925, Clark Papers (SIA), 3:2.

21. Austin H. Clark to Thornton W. Burgess, June 8, 1925, Clark Papers (SIA), 3:2.

22. Thornton W. Burgess to Austin H. Clark, October 1, 1926, Clark Papers (SIA), 3:3.

23. Thornton W. Burgess to Austin H. Clark, June 10, 1925, and Thornton W. Burgess to Austin H. Clark, June 17, 1925, Clark Papers (SIA), 3:2.

24. Thornton W. Burgess to Austin H. Clark, June 17, 1925, Clark Papers (SIA), 3:2.

25. "News Note," *Science* 62 (November 20, 1925): 457.

26. Thornton W. Burgess to Austin H. Clark, January 9, 1928, Clark Papers (SIA), 3:3. In a letter responding to Burgess, Clark said that he had suggested that WEEI, the station owned by the *Boston Transcript*, make an overture to Edward Wigglesworth and the Boston Society of Natural History. Austin H. Clark to Thornton W. Burgess, January 12, 1928, Clark Papers (SIA), 3:3.

27. Ralph H. Lutts, *The Nature Fakers: Wildlife, Science and Sentiment* (Golden: Fulcrum Publishing, 1990), 161. See also Lisa Mighetto, *Wild Animals and American Environmental Ethics* (Tucson: University of Arizona Press, 1991), esp. chap. 1.

28. Austin H. Clark to John C. Phillips, September 29, 1925, Clark Papers (SIA), 9:9.

29. Austin H. Clark to J. R. Schramm, February 12, 1926, Clark Papers (SIA), 10:6.

30. Austin H. Clark to Edward Wigglesworth, October 15, 1925, Clark Papers (SIA), 13:1.

31. Thornton W. Burgess to Austin H. Clark, January 4, 1926, and October 1, 1926, Clark Papers (SIA), 3:3.

32. Thornton W. Burgess to Austin H. Clark, July 10, 1925, Clark Papers (SIA), 3:2.

33. Thornton W. Burgess to Austin H. Clark, September 26, 1925, Clark Papers (SIA), 3:2.

34. Thornton W. Burgess to Austin H. Clark, March 19, 1926, Clark Papers (SIA), 3:3.

35. Arthur Newton Pack and E. Laurence Palmer, *The Nature Almanac: A Handbook of Nature Education* (Washington, D.C.: American Nature Association, 1927), 39.

36. Thornton W. Burgess to Austin H. Clark, March 27, 1925, Clark Papers (SIA), 3:3. Even without the energetic scouts, the caterpillar population may have declined naturally; both the eastern tent and forest tent caterpillars, which are native to New England, emerge in periodic outbreaks lasting three to seven years. U.S. Forest Service, *Forest Insect and Disease Leaflet 9*, http://www.na.fs.fed.us/spfo/pubs/fidls/ftc/tentcat.htm.

37. Thornton W. Burgess to Austin H. Clark, March 28, 1929, Clark Papers (NMNH).

38. Thornton W. Burgess to Austin H. Clark, January 30, 1926, Clark Papers (SIA), 3:3.

39. Austin H. Clark to John C. Merriam, Clark Papers (SIA), 8:6.

40. Thornton W. Burgess to Austin H. Clark, March 19, 1926, Clark Papers (SIA), 3:3.

41. Austin H. Clark to Paul S. Redington, February 23, 1927, Clark Papers (SIA), 10:3; Austin H. Clark to John C. Merriam, February 1, 1926, Clark Papers (SIA), 8:6.

42. Thornton W. Burgess to Austin H. Clark, September 21, 1925, Clark Papers (SIA), 3:2; Austin H. Clark to C. Vaney, March 23, 1927, Clark Papers (SIA), 12:3.

43. John C. Merriam to Austin H. Clark, February 2, 1926, Merriam Papers, box 43, folder "Clark, A. H."; John C. Phillips, "An Investigation of the Periodic Fluctuations in the Numbers of the Ruffed Grouse," *Science* 63 (January 22, 1926): 92–93. See also Thornton W. Burgess to

Austin H. Clark, September 25, 1925, and Austin H. Clark to Thornton W. Burgess, October 25, 1925, Clark Papers (SIA), 3:2; Austin H. Clark to John C. Phillips, October 30, 1925, and John C. Phillips to Austin H. Clark, November 4, 1925, Clark Papers (SIA), 9:9.

44. Austin H. Clark to Thornton W. Burgess, October 25, 1925, Clark Papers (SIA), 3:2; Austin H. Clark to L. O. Howard, March 26, 1926, Clark Papers (SIA), 7:1; Austin H. Clark to Alfred V. Kidder, February 9, 1926, Clark Papers (SIA), 7:8.

45. Thornton W. Burgess to Austin H. Clark, November 8, 1929, Clark Papers (NMNH).

46. Austin H. Clark to Thornton W. Burgess, June 8, 1925, Clark Papers (SIA), 3:2.

47. Ibid.

48. Austin H. Clark to William M. Mann, October 25, 1925, Clark Papers (SIA), 8:5.

49. Ibid.

50. Austin H. Clark to Charles W. Corbett, May 10, 1926, Clark Papers (SIA), 4:4.

51. Carl H. Getz to Austin H. Clark, March 18, 1926, Clark Papers (SIA), 6:1.

52. Austin H. Clark to Carl H. Getz, March 18, 1926, Clark Papers (SIA), 6:1.

53. Austin H. Clark to William M. Mann, July 20, 1926, Clark Papers (SIA), 8:5.

54. Austin H. Clark to Thornton W. Burgess, July 29, 1926, Clark Papers (SIA), 3:3.

55. Shapley attended a Science Service planning meeting in December 1920. See telegram to Vernon Kellogg, December 1, 1920, Hale Papers, roll 65. Shapley later served as Science Service trustee and president.

56. Thornton W. Burgess to Austin H. Clark, September 25, 1925, and September 26, 1925, Clark Papers (SIA), 3:2.

57. Harlow Shapley and Cecilia H. Payne, eds., *The Universe of Stars: Radio Talks from the Harvard Observatory* (Cambridge: Observatory, 1926), 186.

58. Harlow Shapley to Austin H. Clark, January 27, 1926, Clark Papers (SIA), 10:11; "Radio Talks from the Harvard College Observatory," *Science* 62 (November 13, 1925), 431.

59. Harlow Shapley to Austin H. Clark, January 27, 1926, Clark Papers (SIA), 10:11.

60. Ibid.

61. Harlow Shapley to Austin H. Clark, January 20, 1926, and January 25, 1926, Clark Papers (SIA), 10:11.

62. C. G. Abbot to Austin H. Clark, August 25, 1925, Alexander Wetmore Papers, 12:2, Smithsonian Institution Archives.

63. In 1923, the Smithsonian hired a public relations firm to design a new fund-raising campaign. Pamela Henson, Historian of the Smithsonian Institution, personal communication to author, 2004; Ellis L. Yochelson, *Smithsonian Institution Secretary, Charles Doolittle Walcott* (Kent, Ohio: Kent State University Press, 2001). See also Ronald C. Tobey, *The American Ideology of National Science, 1919-1930* (Pittsburgh: University of Pittsburgh Press, 1971), esp. chap. 7; and Daniel J. Kevles, *The Physicists: The History of a Scientific Community in Modern America* (New York: Alfred A. Knopf, 1978), esp. chap. 13.

64. Austin H. Clark to John C. Merriam, February 13, 1927, Merriam Papers, box 43; David Dietz to Austin H. Clark, February 12, 1927, Clark Papers (SIA), 4:9; Thornton W. Burgess to Austin H. Clark, March 8, 1927, Clark Papers (SIA), 3:3.

65. Thornton W. Burgess to Austin H. Clark, August 15, 1925, Clark Papers (SIA), 3:2.

CHAPTER THREE

1. H. L. Smithton to William E. Ritter, January 2, 1924, Science Service Records, 23:6.

2. E. E. Slosson, "Report to the Board of Trustees of Science Service," April 26, 1923, Merriam Papers, 194:1.

3. M. B. R. [Mary Bailey Ritter], "How E. W. and W. E. Met," n.d., Science Service Records, 1920s-1970s, Accession 90-105, 18:24, Smithsonian Institution Archives.

4. Gerald J. Baldasty, *E. W. Scripps and the Business of Newspapers* (Urbana: University of Illinois Press, 1999), 18.

5. Vance H. Trimble, *The Astonishing Mr. Scripps: The Turbulent Life of America's Penny Press Lord* (Ames: Iowa State University Press, 1992), 224-25.

6. William E. Ritter, "Science for the Millions," *Scientific American Monthly* 3 (May 1921): 462.

7. Baldasty, *E. W. Scripps*.

8. James C. Foust, "E. W. Scripps and the Science Service," *Journalism History* 21 (Summer 1995): 60; Frank E. A. Thone and Edna Watson Bailey, "William Emerson Ritter: Builder," *Science* 24 (March 1927): 256-62; Ritter, "Science for the Millions." Ritter and Scripps sought advice and support from powerful scientists like Robert A. Millikan, George Ellery Hale, Vernon Kellogg, and A. A. Noyes. Millikan later claimed that Scripps had to be persuaded to focus exclusively on science rather than economics and government. Robert A. Millikan to Watson Davis, October 31, 1949, Millikan Collection, reel 16, folder 15.4.

9. William E. Ritter to George Ellery Hale, March 1, 1920, Hale Papers, roll 65. See also Ritter, "Science for the Millions."

10. T. D. A. Cockerell, "Why Does Our Public Fail to Support Research?" *Scientific Monthly* 10 (April 1920): 368.

11. Ibid., 371.

12. T. Jackson Lears, *Fables of Abundance: A Cultural History of Advertising in America* (New York: Basic Books, 1994).

13. David Rhees, "The Chemical Foundation and Popular Chemistry between the Wars," *Center for the History of Chemistry (CHOC) News* 3 (Spring 1985): 2; David Rhees, "The Chemists' War: The Impact of World War I on the American Chemical Profession," *Bulletin of the History of Chemistry* 13-14 (1992-1993): 46.

14. Rhees, "Chemists' War," 45; Austin H. Clark, "Science and the Press," *Science* 68 (August 3, 1928): 93. See also George Robert Ehrhardt, "Descendants of Prometheus: Popular Science Writing in the United States, 1915-1948" (Ph.D. diss., Duke University, 1993).

15. Sally Gregory Kohlstedt, Michael M. Sokal, and Bruce V. Lewenstein, *The Establishment of Science in America* (New Brunswick: Rutgers University Press, 1999).

16. Burton E. Livingston to Austin H. Clark, October 15, 1924, Clark Papers (SIA), 8:2.

17. John C. Burnham, *How Superstition Won and Science Lost: Popularizing Science and Health in the United States* (New Brunswick: Rutgers University Press, 1987), esp. 174.

18. Marcel C. LaFollette, *Making Science Our Own: Public Images of Science, 1910-1955* (Chicago: University of Chicago Press, 1990), figs. 2.1 and 2.5.

19. Burnham, *How Superstition Won*; Ronald C. Tobey, *The American Ideology of National Science, 1919-1930* (Pittsburgh: University of Pittsburgh Press, 1971); Daniel J. Kevles, *The Physicists: The History of a Scientific Community in Modern America* (New York: Alfred A. Knopf, 1978), esp. chaps. 12-13.

20. See correspondence between W. J. Humphreys and George Ellery Hale in Hale Papers, reels 65 and 66. The committee was also disbanded and reconstituted during this time.

21. Baldasty, *E. W. Scripps*, 2 and passim.

22. E. W. Scripps to E. E. Slosson, August 1, 1921, Science Service Records, 12:2.

23. Ibid.

24. The best statement of what Scripps believed the organization should be is "Document A—The American Society for the Dissemination of Science," dictated by E. W. Scripps, March 5, 1919, Science Service Records, 1:1. See also William E. Ritter, "The Relation of E. W. Scripps to Science," *Science* 65 (March 25, 1927): 292.

25. When Scripps died in 1926, his will set up a trust account that provided for $30,000 payments annually, with the provision that upon the trust's dissolution, Science Service would receive a lump payment of approximately $500,000. The organization was "sufficiently endowed

to be independent" and yet "prohibited by its charter from making profits." E. E. Slosson to R. S. McBride, April 5, 1921, Science Service Records, 9:6.

26. William E. Ritter to E. W. Scripps, May 13, 1921, Science Service Records, 1:3.

27. E. W. Scripps to E. E. Slosson, August 1, 1921, Science Service Records, 12:2.

28. William E. Ritter to Charles B. Davenport, December 10, 1920, James McKeen Cattell Papers, 184:9, Manuscript Division, Library of Congress.

29. Ibid.

30. "Minutes of the Meeting of the Science News Service held at Miramar, July 7, 1920," and E. W. Scripps to E. E. Slosson, August 28, 1920, Hale Papers, reel 65. See also E. E. Slosson to Vernon Kellogg, July 13, 1920, Central Policy Files, 1919–1923, Science Service folder, National Academies Archives.

31. E. E. Slosson to William E. Ritter, February 7, 1921, Science Service Records, 1:6.

32. E. E. Slosson to Thomas T. Coke, February 7, 1921, Science Service Records, 7:1.

33. Slosson, "New Agency," 323. Slosson initially promised some stringers as much as ten dollars for a five-hundred-word article but changed quickly to five dollars, and most contributors received less. In September 1921, Science Service was paying authors five to six cents per word and then selling those articles for seven cents per word. E. E. Slosson to E. W. Scripps, September 2, 1921, Science Service Records, 12:2. Payments to contributors to the February 1922 *Science News Bulletin* included, for example, four dollars to astronomer Harlow Shapley, four payments, each under six dollars, to astronomer Isabel Lewis, and $1.95 to future staff member Frank Thone.

34. See correspondence between E. E. Slosson and George Washington Carver, March 1921 and July 1921, in Science Service Records, 7:1.

35. E. E. Slosson to L. N. Flint, March 30, 1921, Science Service Records, 69:6.

36. E. E. Slosson to Watson Davis, February 24, 1921, Science Service Records, 7:4.

37. E. E. Slosson to D. T. MacDougal, September 1, 1921, Science Service Records, 9:6.

38. Harry L. Smithton to William E. Ritter, January 3, 1924, Science Service Records, 23:6. Smithton called "Wheeler's exodus from Science Service . . . one of the best things that happened." H. L. Smithton to William E. Ritter, January 4, 1924, Science Service Records, 23:6.

39. E. W. Scripps (on board his yacht) to E. E. Slosson, September 4, 1921, Science Service Records, 12:2.

40. Ibid.

41. William E. Ritter to E. W. Scripps, May 13, 1921, Merriam Papers, box 194, folder "Science News Service." See also William E. Ritter to George Ellery Hale, March 1, 1920, Hale Papers, reel 65.

42. E. E. Slosson to Charles B. Driscoll, April 25, 1921, Science Service Records, 13:6.

43. E. E. Slosson to Frank Richardson Kent, May 4, 1923, Science Service Records, 18:7.

44. See E. E. Slosson to William E. Ritter, December 8, 1921, Science Service Records, 1:6; W. A. Noyes to E. E. Slosson, March 10, 1923, and E. E. Slosson to W. A. Noyes, March 21, 1923, Science Service Records, 20:1.

45. E. E. Slosson to E. W. Scripps, September 2, 1921, Science Service Records, 12:2.

46. Handwritten postscript on letter from E. E. Slosson to D. T. MacDougal, January 5, 1922, Science Service Files, folder "SD, Trees, Annual Rings," Agriculture Collection, National Museum of American History, Smithsonian Institution.

47. "Notes of Talk to Trustees of Science Service by Edwin E. Slosson at the Annual Meeting April 27, 1922," Science Service Records, 1:9.

48. Undated memoranda, circa 1922, from Ralph F. Couch to Howard Wheeler and to Watson Davis, Science Service Records, 55:18.

49. E. E. Slosson to W. A. Cannon, February 23, 1923, Science Service Records, 16:1.

50. E. E. Slosson to Edwin B. Frost, March 12, 1923, Science Service Records, 17:3.

51. E. E. Slosson to Walter S. Adams, August 7, 1923, Science Service Records, 17:3.

52. "To the Managing Editor" circular, August 27, 1923, Science Service Records, 62:1.

53. H. L. Smithton to W. E. Ritter, January 2, 1924, Science Service Records, 23:6.

54. Ibid.

55. As quoted, ibid.

56. Ibid.

57. Ibid.

CHAPTER FOUR

1. E. E. Slosson to William E. Ritter, February 7, 1921, Science Service Records, 1:6.

2. "A Brief Sketch of 'Make It Yourself'" and "To the Managing Editor," advertising circulars dated January 16, 1922, Science Service Records, 60:5; "Statement of Managing Editor to Board of Trustees of Science Service," April 26, 1923, Merriam Papers, 194:1. Final expenses for the project were $8,919.47; income, $7,110.24.

3. "Report of Manager of Science Service," undated [circa June–December 1921], Science Service Records, 1:2.

4. E. E. Slosson to Ida Clyde Clarke, September 2, 1925, Science Service Records, 25:8.

5. E. E. Slosson to John Mills, March 17, 1924, Science Service Records, 23:1.

6. E. E. Slosson to Dayton C. Miller, February 25, 1925, Science Service Records, 27:3.

7. E. E. Slosson to Alice Templin Rankin, February 3, 1925, Science Service Records, 29:10.

8. E. E. Slosson to Austin H. Clark, May 20, 1924, Clark Papers (SIA), 10:11.

9. Austin H. Clark to Walter M. Gilbert, May 22, 1924, Merriam Papers, box 43, folder "Clark, A. H."

10. E. E. Slosson to George R. Mansfield, May 31, 1924, Science Service Records, 23:1.

11. Ibid.

12. E. E. Slosson to H. E. Howe, June 17, 1924, Science Service Records, 22:6.

13. E. E. Slosson to George R. Mansfield, June 4, 1924, Science Service Records, 23:1.

14. E. E. Slosson to George R. Mansfield, June 21, 1924, Science Service Records, 23:1.

15. Frank Thone to Carroll Lane Fenton, August 3, 1942, Science Service Records, 234:11.

16. Frank Thone to Frank McDonough, July 29, 1940, Science Service Records, 220:2.

17. Frank Thone to E. E. Slosson, January 7, 1924, Science Service Records, 34:5.

18. Frank Thone to Alice L. Braunwarth Halstead, August 9, 1932, Science Service Records, 136:1.

19. Frank Thone to E. E. Slosson, February 12, 1924, Science Service Records, 19:7.

20. E. E. Slosson to Frank Thone, November 14, 1923, Science Service Records, 19:7.

21. E. E. Slosson, "Memorandum to Dr. Tisdale," June 23, 1924, Science Service Records, 24:1.

22. "Science News of the Week," typed script for broadcast on WCAP, July 23, 1924.

23. Typewritten draft for "Radio Talk Aug. 20," with handwritten additions by Watson Davis, Science Service Records, 103:2. Subsequent quotes from same draft.

24. Maurice Holland to F. B. Jewett, November 24, 1924, Science Service Records, 89:2.

25. A. S. Barrows to Watson Davis, April 26, 1925, Science Service Records, 118:6.

26. A. S. Barrows to Watson Davis, March 4, 1926, Science Service Records, 118:6.

27. Watson Davis to Albert Barrows, April 9, 1926, Central Policy Files, 1924-1931, Executive Board, Committee on Radio Talks, National Academies Archives.

28. E. E. Slosson to W. L. Chenery, June 20, 1925, Science Service Records, 90:8.

29. Davis said that the "very publicity shy" Einstein "would not really allow us to sit down. He refused absolutely to write [an article for Science Service] and the circumstances were not auspicious for asking him about the future of physics." Watson Davis to E. E. Slosson, September 20, 1925, Science Service Records, 28:2.

30. See Marcel Chotkowski LaFollette, *Reframing Scopes: Journalists, Scientists, and Lost Photographs from the Trial of the Century* (Lawrence: University Press of Kansas, 2008).

31. "Will Bring Evolution Case to Your Home," *Chicago Tribune*, June 28, 1925; "Broadcast of Scopes Trial Unprecedented," *Chicago Daily Tribune*, July 5, 1925; *WGN: A Pictorial History* (Chicago: WGN, 1961), 21–22; James Walter Wesolowski, "Before Canon 35: WGN Broadcasts the Monkey Trial," *Journalism History* 2 (Fall 1975): 76–79, 86–87; Erik Barnouw, *A Tower in Babel: A History of Broadcasting in the United States to 1933* (New York: Oxford University Press, 1966), 196–97; Donald G. Godfrey and Frederic A. Leigh, eds., *Historical Dictionary of American Radio* (New York: Greenwood Press, 1998), 359.

32. "Evolution Sidelights," *Atlanta Constitution*, July 12, 1925.

33. "Broadcast of Scopes Trial Unprecedented."

34. "Inside the Loud Speaker with Quin A. Ryan, the Voice of W-G-N," *Chicago Daily Tribune*, July 26, 1925.

35. Elmer Douglass, "Elmer Haunts Scopes' Trial Radio Waves," *Chicago Daily Tribune*, July 16, 1925.

36. Watson Davis to James Stokley, January 19, 1925, Science Service Records, 81:8.

37. Watson Davis to William E. Ritter, January 7, 1925, Science Service Records, 81:5.

38. "Women of Science Service," a Smithsonian Institution Archives Web exhibit, summarizes the lives of some of these pioneering women writers (http://siarchives.si.edu/research/sciservwomen.html).

39. Clients included stations owned by Ohio State University, Rensselaer Polytechnic Institute, Antioch College, Rollins College, the *Chicago Daily News*, and the *Boston Transcript*.

40. E. E. Slosson to Sir Robert Robertson, October 8, 1926, Science Service Records, 33:7.

41. Albert L. Barrows to Watson Davis, October 18, 1926, Science Service Records, 89:2.

42. "Report of Committee on Radio Talks, 1925–1926," National Research Council, October 11, 1926, Science Service Records, 89:2.

43. Austin H. Clark to James McKeen Cattell, November 6, 1926, Clark Papers (SIA), 3:6; Austin H. Clark to Edward Wigglesworth, November 9, 1926, Clark Papers (SIA), 13:1.

44. Austin H. Clark to Edward Wigglesworth, November 9, 1926, Clark Papers (SIA), 13:1.

45. Austin H. Clark to Thornton W. Burgess, November 15, 1926, Clark Papers (SIA), 3:2.

46. Ibid.

47. James Stokley to Edward B. Husing, April 13, 1927, Science Service Records, 89:2.

48. James Stokley to Austin H. Clark, December 8, 1926, and Austin H. Clark to James Stokley, December 11, 1926, Science Service Records, 89:2.

49. Station WOO's program director, Harriette G. Ridley, was attempting to schedule and promote the talks and was frantically writing (and wiring) for titles and names of speakers while Stokley and Clark were squabbling. See correspondence between James Stokley and Harriette G. Ridley, Science Service Records, 89:2.

50. Form letter sent to editors from Watson Davis, September 11, 1922, Science Service Records, 60:2.

51. "Notes of a Talk to Trustees of Science Service at the meeting of June 17, 1921," Science Service Records, 1:2.

52. Ibid.

53. Ibid.

54. Ibid.

55. Edward L. Bernays, "Manipulating Public Opinion: The Why and the How," *American Journal of Sociology* 33 (May 1928): 958–71.

56. Austin H. Clark to Thornton W. Burgess, June 1, 1925, Clark Papers (SIA), 3:2.

57. C. G. Abbot to E. E. Slosson, September 10, 1925, Science Service Records, 27:11.

58. Burton E. Livingston to Watson Davis, October 27, 1927, Science Service Records, 91:9.

59. Frank Thone to Clarence H. Kennedy, December 1, 1928, Science Service Records, 96:6.

60. Philip J. Sinnott to Marlen Pew, circa October 4, 1927, enclosed with letter from Marlen Pew to E. E. Slosson, October 4, 1927, Science Service Records, 443:9.

61. William E. Ritter to Watson Davis, November 10, 1927, Science Service Records, 443:9.

62. E. E. Slosson, "The Journalist as Middleman in Science," January 14, 1926, Science Service Records, 48:3.

63. Ibid.

CHAPTER FIVE

1. Harry L. Smithton to Watson Davis, September 27, 1927, Science Service Records, 120:2.

2. Sydney W. Head, *Broadcasting in America: Survey of Television and Radio* (Cambridge: Riverside Press, 1956), 400.

3. Thomas Streeter, *Selling the Air: A Critique of the Policy of Commercial Broadcasting in the United States* (Chicago: University of Chicago Press, 1996), 87.

4. Smithsonian Institution, *Report on the Progress and Condition of the United States National Museum for the Year Ended June 30, 1927* (Washington, D.C.: U.S. Government Printing Office, 1927), 18. See also *Science* 64 (November 26, 1926): 520.

5. Streeter, *Selling the Air.*

6. Mark Goodman, "The Radio Act of 1927 as a Product of Progressivism," *Media History Monographs* 2 (1998-1999): 3, citing *Congressional Record* (1927), 5478. By the end of 1927, the number of amateur stations exceeded eighteen thousand.

7. E. E. Slosson to C. D. Wagoner, General Electric Company News Bureau, February 5, 1925, Science Service Records, 26:4.

8. See Erik Barnouw, *A Tower in Babel: A History of Broadcasting in the United States to 1933* (New York: Oxford University Press, 1966); Erik Barnouw, *The Golden Web: A History of Broadcasting in the United States, 1933-1953* (New York: Oxford University Press, 1968); Michelle Hilmes, *Radio Voices: American Broadcasting, 1922-1952* (Minneapolis: University of Minnesota Press, 1997); Streeter, *Selling the Air.*

9. Goodman, "Radio Act," 5.

10. Ibid.

11. This is analogous to communication via the Internet, where someone can, in theory, transmit any words, no matter how offensive, inflammatory, or dangerous, but internet service providers (ISPs) and gateway institutions (universities or companies) can regulate their own users and thereby censor certain speech by prohibiting access through their connections. Attempts have also been made to prosecute these unofficial regulators, after the fact, for crimes committed by way of the connections they provided, such as using the Internet to solicit minors. Much of the debate over Internet speech has revolved around whether it should or should not be regulated at all and, if so, by whom and at what stage in transmission.

12. For discussion of the development of British radio, see Andrew Crisell, *Understanding Radio* (London: Methuen, 1986); Anthony Smith, *The Shadow in the Cave: The Broadcaster, His Audience, and the State* (Urbana: University of Illinois Press, 1973); and Anthony Smith, ed., *Television: An International History* (New York: Oxford University Press, 1995).

13. Robert E. Summers, ed., *Wartime Censorship of Press and Radio* (New York: H. W. Wilson, 1942); Michael S. Sweeney, *Secrets of Victory: The Office of Censorship and the American Press and Radio in World War II* (Durham: University of North Carolina Press, 2001). From time to time, certain words have been ruled off-limits and advertising of certain products restricted. Public protest has also, on occasion, prompted station and network owners to include or exclude content but not to any greater extent than with other types of mass communication.

14. Streeter, *Selling the Air,* 98.

15. President's Research Committee on Social Trends, *Recent Social Trends in the United States,* vol. 2 (New York: McGraw-Hill, 1933), 942; Susan J. Douglas, *Listening In: Radio and the American Imagination* (New York: Times Books, 1999), 131, citing data from Hadley Cantril and Gordon Allport. See also Hilmes, *Radio Voices.*

16. Douglas, *Listening In,* 76; Hilmes, *Radio Voices.*

17. Streeter, *Selling the Air,* 98-99.

18. Douglas B. Craig, *Fireside Politics: Radio and Political Culture in the United States, 1920-1940* (Baltimore: Johns Hopkins University Press, 2000), 34.

19. Craig, *Fireside Politics,* 25. See also Susan Smulyan, *Selling Radio: The Commercialization of American Broadcasting, 1920-1934* (Washington, D.C.: Smithsonian Institution Press, 1994); Erik Barnouw, *The Sponsor: Notes on a Modern Potentate* (New York: Oxford University Press, 1978).

20. Harlow Shapley to Austin H. Clark, January 20, 1926, and Austin H. Clark to Harlow Shapley, January 23, 1926, Clark Papers (SIA), 10:11.

21. Streeter, *Selling the Air,* 98.

22. Craig, *Fireside Politics,* 25.

23. National Broadcasting Company, "Note to Radio Editors, Rate Card," December 16, 1927, Science Service Records, 118:6.

24. Craig, *Fireside Politics,* 25; Robert W. McChesney, "Media and Democracy: The Emergence of Commercial Broadcasting in the United States, 1927-1935," *OAH Magazine of History* 6 (Spring 1992), citing 1935 publications.

25. Streeter, *Selling the Air,* 102-3.

26. Judith C. Waller, *Broadcasting in the Public Service* (Chicago: John C. Swift, 1943).

27. Roland Marchand, *Advertising the American Dream: Making Way for Modernity, 1920-1940* (Berkeley: University of California Press, 1985), 88-90, 109.

28. Streeter, *Selling the Air,* 108.

29. Christopher H. Sterling and John M. Kittross, *Stay Tuned: A Concise History of American Broadcasting* (Belmont: Wadsworth, 1978), 78.

30. Head, *Broadcasting in America,* 400.

31. Craig, *Fireside Politics,* 68. Craig says that 164 of 202 early stations eventually folded or sold out to commercial interests. See also Barnouw, *Sponsor,* 27-28.

32. Mitchell V. Charnley, *News by Radio* (New York: Macmillan, 1948), 8.

33. Austin H. Clark to Thornton W. Burgess, September 28, 1926, Clark Papers (SIA), 3:3.

34. Austin H. Clark to Thornton W. Burgess, March 21, 1928, Clark Papers (SIA), 3:3.

35. Thornton W. Burgess to Austin H. Clark, October 1, 1926, Clark Papers (SIA), 3:3.

36. Austin H. Clark to Thornton W. Burgess, March 21, 1928, Clark Papers (SIA), 3:3.

37. Thornton W. Burgess to Austin H. Clark, June 17, 1925, Clark Papers (SIA), 3:2.

38. Thornton W. Burgess to Austin H. Clark, April 30, 1928, Clark Papers (SIA), 3:3. Burgess did not identify that "angel" to Clark.

39. Thornton W. Burgess to Austin H. Clark, August 16, 1929, Clark Papers (NMNH).

40. Thornton W. Burgess to Austin H. Clark, October 18, 1929, Clark Papers (NMNH).

41. Thornton W. Burgess to Austin H. Clark, November 8, 1929, Clark Papers (NMNH).

42. Thornton W. Burgess to Austin H. Clark, November 26, 1929, Clark Papers (NMNH).

43. Ibid.

44. Thornton W. Burgess to Austin H. Clark, March 15, 1930, Clark Papers (NMNH).

45. Thornton W. Burgess to Austin H. Clark, February 5, 1930, Clark Papers (NMNH).

46. E. E. Slosson, "Annual Address of the Director of Science Service at Meeting of the Board of Trustees, Washington, April 25, 1929," Merriam Papers, box 164, folder "Slosson, E. E."

47. James Stokley to H. K. Boice, February 12, 1929, Science Service Records, 102:1.

48. E. E. Slosson, memorandum to James Stokley and Watson Davis, March 25, 1929, Science Service Records, 41:2.

49. Draft scripts for *Science Snapshots*, April 5, 1929, Science Service Records, 112:4.

50. Script for *Science Snapshots*, April 26, 1929, Science Service Records, 112:4.

51. Marcel C. LaFollette, *Making Science Our Own: Public Images of Science, 1910–1955* (Chicago: University of Chicago Press, 1990); Marcel C. LaFollette, "Eyes on the Stars: Images of Women Scientists in Popular Culture," *Science, Technology, & Human Values* 13 (Fall 1988): 262–75. E. E. Slosson's wife, May Preston Slosson, was a well-known suffragette, and Slosson himself had spoken at women's suffrage rallies.

52. James Stokley to Julian F. Seebach, August 12, 1929, and September 4, 1929, and Julian F. Seebach to James Stokley, August 19, 1929, and September 7, 1929, Science Service Records, 112:6.

53. Margaret E. Young to James Stokley, September 11, 1929, Science Service Records, 112:6.

54. James Stokley to William S. Paley, December 30, 1929, Science Service Records, 112:2.

55. Frank Thone to Winterton C. Curtis, July 2, 1930, Science Service Records, 112:3. See also description in "Science Service Radio Talks over Columbia Broadcasting System," circa 1931, Alexander Wetmore Papers, Record Unit 7006, 117:1, Smithsonian Institution Archives.

56. Austin H. Clark to James McKeen Cattell, January 14, 1930, Clark Papers (NMAH). See also Austin H. Clark to Harlow Shapley, January 20, 1930, Clark Papers (NMAH); Austin H. Clark to Robert A. Millikan, January 14, 1930, Robert A. Millikan Collection, reel 18; and Austin H. Clark to John C. Merriam, January 14, 1930, Merriam Papers, box 43. Clark's announcement of the series in *Science* listed these and other scientists but no members of the press. Austin H. Clark, "Radio Talks," *Science* 71 (May 2, 1930): 454.

57. Austin H. Clark to James McKeen Cattell, June 25, 1930, Clark Papers (NMAH).

58. James Stokley to Guy H. Lagroe, October 31, 1930, Science Service Records, 116:1.

59. James Stokley to M. R. Baker, January 28, 1930, Science Service Records, 118:8.

60. Maurice Holland to Watson Davis, August 1930, Science Service Records, 114:11; R. P. Shaw to Watson Davis, July 15, 1930, Science Service Records, 112:1. The NRC and the Science Advisory Council arranged fifteen-minute talks on NBC from November 5, 1930, to February 25, 1931, with speakers like Karl Compton, Arthur D. Little, Fay-Cooper Cole, Isaiah Bowman, and D. T. MacDougal.

61. James McKeen Cattell to Watson Davis, February 16, 1931, Science Service Records, 123:6. The trustees even briefly considered Austin Clark as Slosson's replacement because he was "superior as a scientist to Watson Davis." Edwin B. Wilson to David White, December 24, 1929, Central Policy Files, 1924–1931, Science Service folder, National Academies Archives.

62. In a 1987 interview with historian Bruce Lewenstein, former Science Service writer Jane Stafford stated that the "scientists did not accord to newspaper people or even science writers the status that they accorded other scientists or people in other professions" and that those prejudices played a significant role in the hesitation to appoint Davis. "National Association of Science Writers—Jane Stafford," transcript of an interview conducted by Bruce V. Lewenstein, February 6, 1987, National Association of Science Writers Archives, Cornell University.

63. W. H. Howell to Watson Davis, June 24, 1936, Science Service Records, 4:3.

64. Watson Davis to David White, February 10, 1931, Science Service Records, 131:9.

65. Watson Davis to J. W. Foster, Scripps-Howard Newspapers, February 12, 1931, Science Service Records, 124:10.

66. Emily C. Davis to potential stringers (several letters with same paragraph), October 1931, Science Service Records, 124:3.

67. Frank Thone to potential stringer, January 26, 1932, Science Service Records, 132:1.

68. Frank Thone to Theodor G. Ahrens, February 18, 1932, Science Service Records, 132:4.

69. Watson Davis to William E. Ritter, May 18, 1932, Science Service Records, 140:2.

70. "Report to the Annual Meeting of the Board of Trustees of Science Service, Thursday, April 28, 1932," by Watson Davis, Science Service Records, 3:2.

71. Davis initially estimated the cost at $125 per program (Watson Davis to Walter V. Bingham, March 31, 1932, Science Service Records, 132:11) but revised that figure considerably six months later. "Information Memorandum on Progress of Science Service, October 18, 1932, Science Service Records, 3:3. Speakers were Leo H. Baekeland, Karl T. Compton, William M. Mann, John C. Merriam, Robert Millikan, and William H. Welch.

72. Remarks by Karl T. Compton in "Science Service Conference," *Science* 76 (August 19, 1932): 152.

73. Remarks by A. H. Kirchhofer, ibid., 153.

74. Remarks by Robert P. Scripps, ibid., 156.

75. Frank Thone to Theodor Ahrens, April 4, 1933, Science Service Records, 143:6.

76. Gabriele Rabel to Frank Thone, April 22, 1933, Science Service Records, 149:9.

77. Theodor G. Ahrens to Frank Thone, April 7, 1933, Science Service Records, 143:6.

78. Watson Davis to H. E. Howe, May 25, 1933, Science Service Records, 146:7.

79. Maxim Bing to Watson Davis, June 20, 1933, Science Service Records, 144:1. By 1934 Thone could not always disguise his sorrow and anger at what was happening in Germany. In a lengthy and impolitic response to a subscriber (Frank Thone to W. J. Lenz, September 6, 1934, Science Service Records, 156:2), Thone declared that he intended to speak out often against Nazi pseudoscience: "As a believer in sound scientific investigation . . . and further as a believer and supporter in freedom of speech and freedom of the press, I shall continue to attack with all the weapons at my command (be they of reason or of ridicule) such lunatic proceedings as we are now compelled to witness in the land of my forefathers, under its present thug government."

80. Watson Davis to Maxim Bing, July 1, 1933, Science Service Records, 144:1. See, for example, Selig Hecht to Frank Thone, January 28, 1936 (Science Service Records, 174:8), regarding tuberculosis researcher Richard Scherman.

81. A. M. Sperber, *Murrow: His Life and Times* (New York: Freundlich Books, 1986), chap. 3.

82. Watson Davis to I. I. Rabi, December 1, 1933, Science Service Records, 149:7. Franck's travel to the United States had been arranged through Duggan's Emergency Committee in Aid of Displaced German Scholars.

83. Watson Davis to Otis W. Caldwell, November 12, 1931, Science Service Records, 122:7.

84. Frank Thone to Jay N. Darling, March 13, 1934, Science Service Records, 153:9.

85. Thornton W. Burgess to Austin H. Clark, January 24, 1931, Clark Papers (NMNH).

86. Thornton W. Burgess to Austin H. Clark, February 8, 1935, Clark Papers (SIA), 3:3.

87. "Dear Radio Neighbor," letter from Thornton W. Burgess, circa 1935, Clark Papers (NMNH).

88. *Radio Nature League News* (February 1936), 1.

89. Thornton W. Burgess to Austin H. Clark, September 14, 1935, Clark Papers (NMNH).

90. Watson Davis, *The Advance of Science* (New York: Doubleday Doran, 1934), 375.

CHAPTER SIX

1. Remarks by Maurice Holland in *Educational Broadcasting 1936, Proceedings of the First National Conference on Educational Broadcasting, held in Washington, DC, on December 10, 11, and 12, 1936*, ed. C. S. Marsh (Chicago: University of Chicago Press, 1937), 322.

2. Susan Smulyan, *Selling Radio: The Commercialization of American Broadcasting, 1920-1934* (Washington, D.C.: Smithsonian Institution Press, 1994), 131–42; Paul R. Gorman, *Left Intellectuals and Popular Culture in Twentieth-Century America* (Chapel Hill: University of North Carolina Press, 1996).

3. Douglas B. Craig, *Fireside Politics: Radio and Political Culture in the United States, 1920–1940* (Baltimore: Johns Hopkins University Press, 2000), 213.

4. Robert W. McChesney, "Media and Democracy: The Emergence of Commercial Broadcasting in the United States, 1927–1935," *OAH Magazine of History* 6 (Spring 1992); Joy Elmer Morgan, in "Education's Rights on the Air" (1931), as quoted in Eugene E. Leach, "Snookered 50 Years Ago," *Current*, January, February, and March 1983 (republished online as Eugene E. Leach, "Tuning Out Education: The Cooperative Doctrine in Radio," December 13, 1999, http://www.current.org/coop).

5. Morgan, "Education's Rights on the Air."

6. Smulyan, *Selling Radio*, 140–42. See also Leach, "Tuning Out Education," online version, pt. 3; Sydney W. Head, *Broadcasting in America: Survey of Television and Radio* (Cambridge: Riverside Press, 1956); and Levering Tyson, ed., *Radio and Education, Proceedings of the First National Assembly of the National Advisory Council on Radio in Education, 1931* (Chicago: University of Chicago Press, 1931).

7. According to Leach, the president of the Carnegie Corporation "had the final word in NACRE affairs," while Tyson ran daily operations. Leach, "Tuning Out Education." See Levering Tyson, *Education Tunes In: A Study of Radio Broadcasting in Adult Education* (New York: American Association for Adult Education, 1930).

8. For a summary of Tyson's views, see Leach, "Tuning Out Education," online version, pt. 2.

9. See Levering Tyson to Robert A. Millikan, February 25, 1931, Millikan Collection, reel 17.

10. Robert A. Millikan to Levering Tyson, February 18, 1931, and Levering Tyson to Robert A. Millikan, February 23, 1931, Millikan Collection, reel 17.

11. Levering Tyson to Robert A. Millikan, February 23, 1931, Millikan Collection, reel 17.

12. Tyson, *Education Tunes In*, 26.

13. See Leach, "Tuning Out Education," and McChesney, "Media and Democracy."

14. Robert A. Millikan, *Reprint of His Address in May 1931 to the First National Assembly of the National Advisory Council on Radio in Education* (Chicago: University of Chicago Press, 1931), 5.

15. Millikan's suggestions for science programs were not received enthusiastically. See Levering Tyson to Robert A. Millikan, various dates, Millikan Collection, reel 17.

16. Craig, *Fireside Politics*, 213.

17. Tyson, *Proceedings of the First National Assembly*, 42.

18. Watson Davis, "Report of the Committee on Science," ibid., 258–61.

19. The principle Davis described was that of Raman spectra. "Raman spectra, which constitute light signals from the vibrations of the atoms of chemical molecules, when transposed into sound vibrations of audible frequency, can actually be played on the piano." Ibid., 258–59.

20. Robert A. Millikan, "Radio's Part in the Creation of an Intelligent Electorate," in *Education on the Air . . . and Radio and Education 1935: Proceedings of the Sixth Annual Institute for Education by Radio . . . Combined with the Fifth Annual Assembly of the National Advisory Council on Radio in Education*, ed. Levering Tyson and Josephine MacLatchy (Chicago: University of Chicago Press, 1935), 10–16.

21. Programs ran weekly from October 17, 1931, to May 21, 1932, on forty to forty-five stations nationwide. Records of the American Psychological Association, box 525, folders on "National Advisory Council on Radio in Education, 1930–1937," Manuscript Division, Library of Congress; Tyson, *Proceedings of the First National Assembly*; Walter V. Bingham, "An Experiment in Broadcasting Psychology: Report of the Committee on Psychology," in *Radio and Education, Proceedings of the Second Assembly of the National Advisory Council on Radio in Education, 1932*, ed. Levering Tyson (Chicago: University of Chicago Press, 1932), 27; Walter V. Bingham, ed., *Psychology Today: Lectures and Study Manual* (Chicago: University of Chicago Press, 1932).

22. Bingham, "Experiment in Broadcasting Psychology," 28.

23. "National Advisory Council on Radio in Education Interim Report of the Committee on Psychology," August 3, 1931, Records of the American Psychological Association, box 525, folder

"National Advisory Council on Radio in Education, 1930-1937—Reports," Manuscript Division, Library of Congress.

24. National Advisory Council on Radio in Education, *Psychology Today*, Listener Notebook no. 1, prepared by Henry E. Garrett and Walter V. Bingham (1931), 28, copy in Science Service Records, 140:5.

25. Head, *Broadcasting in America*, 402-3.

26. Ibid., 401. The Federal Radio Education Committee, established in 1935, also could not persuade commercial broadcasters to increase public service programming.

27. For accounts of Brinkley's radio career, see Craig, *Fireside Politics*, 73-74; Gene Fowler and Bill Crawford, *Border Radio* (Austin: Texas Monthly Press, 1987); Donald G. Godfrey and Frederic A Leigh, eds., *Historical Dictionary of American Radio* (New York: Greenwood Press, 1998), 53; Michelle Hilmes, *Only Connect: A Cultural History of Broadcasting in the United States* (Belmont: Wadsworth, 2001), 11-15; Tom Lewis, *Empire of the Air: The Men Who Made Radio* (New York: HarperCollins, 1991), 237; and R. Alton Lee, *The Bizarre Careers of John R. Brinkley* (Lexington: University Press of Kentucky, 2002).

28. Jon D. Swartz and Robert C. Reinehr, *Handbook of Old-Time Radio: A Comprehensive Guide to Golden Age Radio Listening and Collecting* (Metuchen, N.J.: Scarecrow Press, 1993), 166.

29. Copeland was the Democratic senator from New York from 1923 to 1938. His fifteen-minute programs were broadcast four to five mornings a week on NBC-Blue (1927-1930), NBC (1931-1932), and CBS (1932). Prior to his election, Copeland taught ophthalmology and otology at the University of Michigan Medical School and served as New York City Commissioner of Public Health.

30. Iago Goldston in New York Academy of Medicine, *Radio in Health Education* (New York: Columbia University Press, 1945). See also, in the same volume, Ernest L. Stebbins, "The Responsibility of the Radio Industry in Public Health."

31. President's Research Committee on Social Trends, *Recent Social Trends in the United States*, vol. 2, (New York: McGraw-Hill, 1933), 883; W. W. Bauer and Thomas G. Hull, *Health Education of the Public: A Practical Manual of Technic* (Philadelphia: W. B. Saunders, 1937), 41-42.

32. Winfred Wylam Bird, *An Analysis of the Aims and Practice of the Principal Sponsors of Education by Radio in the United States*, University of Washington Extension Series Bulletin no. 10 (August 1939), 67. Bauer and Hall, *Health Education*, 40-41. Elizabeth Toon argues that these commercial venues had the adverse effect of sending a message that personal health advice was a commodity. Elizabeth A. Toon, "Managing the Conduct of the Individual Life: Public Health Education and American Public Health, 1910 to 1940" (Ph.D. diss., University of Pennsylvania, 1998).

33. Bauer and Hall, *Health Education*, 43-44.

34. Austin H. Clark to Albert P. Taylor, September 16, 1924, Clark Papers (SIA), 12:1.

35. Smulyan, *Selling Radio*, 14-20, 96.

36. Austin H. Clark to Major-General A. W. Greely, October 8, 1924, Clark Papers (SIA), 6:3.

37. Ibid.

38. Austin H. Clark to Edward M. Crane, October 11, 1924, Clark Papers (SIA), 4:6. See also Austin H. Clark to Col. E. Lester Jones, December 1, 1924, Clark Papers (SIA), 7:6.

39. Austin H. Clark to E. A. Back, March 24, 1926, and E. A. Back to Austin H. Clark, March 26, 1926, Clark Papers (SIA), 1:5. Back said that he had received many telephone calls and sixty-eight letters.

40. *Annual Report of the Board of Regents of the Smithsonian Institution, 1925* (Washington, D.C.: U.S. Government Printing Office, 1926), 12; *Annual Report of the Board of Regents of the Smithsonian Institution, 1926* (Washington, D.C.: U.S. Government Printing Office, 1927), 16.

41. Austin H. Clark, "Radio Talks," *Scientific Monthly* 35 (October 1932): 353.

42. *Annual Report of the Board of Regents of the Smithsonian Institution, 1927* (Washington, D.C.: U.S. Government Printing Office, 1928), 18.

43. Austin H. Clark to C. D. Walcott, January 7, 1925, SI Secretary (RU46), 15:8. See also various correspondence between James McKeen Cattell and Austin H. Clark in 1925, Clark Papers (SIA), 3:6.

44. Harlow Shapley to Austin H. Clark, January 20, 1926, Clark Papers (SIA), 10:11. See also Harlow Shapley and Cecilia H. Payne, eds., *The Universe of Stars: Radio Talks from the Harvard Observatory* (Cambridge: Observatory, 1926), as well as the 1929 revised edition.

45. Robert T. Hance, *A Series of Eight Radio Talks on Zoology, Old and New (with Select Bibliography)*, Radio Publication no. 39 (University of Pittsburgh, 1928), 3.

46. Watson Davis to R. G. Hoskins, February 12, 1935, Science Service Records, 403:59.

47. Tabulation of 2,035 listener requests for *Science News of the Week* scripts, June 1926–November 1929. Attachment to Watson Davis to H. Robinson Shipherd, February 28, 1930, Science Service Records, 118:5.

48. Robert F. Elder to Watson Davis, October 27, 1934, and Watson Davis to Robert F. Elder, November 2, 1934, Science Service Records, 154:2. Elder was conducting research on how to measure the "degree of listener interest . . . aroused by broadcasts dealing with scientific subjects."

49. Henry Field to Watson Davis, April 30, 1935, Science Service Records, 403:37.

50. Henry Field to Watson Davis, March 29, 1939, Science Service Records, 385:10.

51. NACRE enrollment blank, Science Service Records, 140:6.

52. Walter V. Bingham, "An Experiment in Broadcasting Psychology: Report of the Committee on Psychology," 27. All booklets and lectures were compiled and reprinted as Walter V. Bingham, ed., *Psychology Today*. APA distributed 64,685 copies of the individual lectures.

53. Comparison based on radio listener data in President's Research Committee on Social Trends, *Recent Social Trends in the United States*, vol. 1 (New York: McGraw-Hill, 1933). AAAS mailed 8,000 questionnaires and received 2,000 responses, from which 1,250 were randomly selected for analysis. F. R. Moulton, "Science by Radio," *Scientific Monthly* 47 (December 1938): 546–48.

54. Neville Miller, "The Broadcaster Speaks," *Journal of Educational Sociology* 14 (February 1941): 323–24. Miller was president of the National Association of Broadcasters. See also Davis, "Report of the Committee on Science," 259–60.

55. Austin H. Clark, "Science and the Radio," *Scientific Monthly* 34 (March 1932): 268.

56. Edited transcript of Symposium on Science and the Press held at AAAS Annual Meeting, Pittsburgh, 1934, p. 3, Smithsonian Institution, Pawson/NMNH, Austin H. Clark Papers, Cattell folder.

57. See Benjamin C. Gruenberg, *Science and the Public Mind* (New York: McGraw-Hill, 1935); and Benjamin C. Gruenberg, "Science and the Layman," *Scientific Monthly* 40 (May 1935): 450–57.

58. Austin H. Clark, "Selling Entomology," *Scientific Monthly* 32 (June 1931): 534. See also Austin H. Clark, "Radio Talks," *Scientific Monthly* 35 (October 1932): 352–59.

59. W. F. Austin to C. G. Abbot, September 9, 1932, SI Secretary (RU46), 87:3.

60. Austin H. Clark to Thornton W. Burgess, June 20, 1925, Clark Papers (SIA), 3:2.

61. Austin H. Clark, "Science and the Radio," *Scientific Monthly* 34 (March 1932): 268–72. In the draft of the essay, Clark wrote that science was too often interpreted "in terms of dollars" or as a "fascinating mystery." Austin H. Clark, "Museums and the Radio," undated draft, pp. 1–2, Clark Papers (SIA), 16:5.

62. Austin H. Clark, "Museums and the Radio," undated draft, p. 4, Clark Papers (SIA), 16:5.

63. Austin H. Clark, comments on a draft manuscript, undated, Clark Papers (SIA), 16:5.

64. J. S. Ames to William H. Howell, May 21, 1934, Science Service Records, 151:9.

65. Remarks by Maurice Holland, in Marsh, *Educational Broadcasting 1936*, 321–22.

66. Ibid., 322.

67. Ibid.

68. Ibid., 323.

CHAPTER SEVEN

1. Max Wylie, *Best Broadcasts of 1939–40* (New York: Whittlesey House, 1940), 263.

2. Douglas B. Craig, *Fireside Politics: Radio and Political Culture in the United States, 1920–1940* (Baltimore: Johns Hopkins University Press, 2000).

3. Watson Davis, "Report of the Committee on Science," in Levering Tyson, ed., *Radio and Education, Proceedings of the First National Assembly of the National Advisory Council on Radio in Education, 1931* (Chicago: University of Chicago Press, 1931), 260.

4. "Minutes of the Meeting of the Executive Committee of Science Service . . . March 30, 1935," p. 4, Science Service Records, 3:8.

5. G. Emerson Markham to Watson Davis, February 8, 1939, Science Service Records, 209: 2.

6. Neil B. Reynolds and Ellis L. Manning, ed., *Excursions in Science* (New York: McGraw-Hill, 1939), viii. See also General Electric Company News Bureau, press release, Science Service Records, 383:18; Comments by G. Emerson Markham in C. S. Marsh, ed., *Educational Broadcasting 1936, Proceedings of the First National Conference on Educational Broadcasting, held in Washington, DC, on December 10, 11, and 12, 1936* (Chicago: University of Chicago Press, 1937), 323; James Stokley, ed., *Science Marches On* (New York: Ives Washburn, 1951); Neil B. Reynolds and Ellis L. Manning, eds., *Excursions in Science* (Freeport: Books for Libraries Press, 1972).

7. Barnouw calls the DuPont project an exercise in "image-repair." Erik Barnouw, *The Sponsor: Notes on a Modern Potentate* (New York: Oxford University Press, 1978), 34. Initial investments in the 1935 public relations campaign exceeded half a million dollars. William L. Bird Jr., *"Better Living": Advertising, Media, and the New Vocabulary of Business Leadership, 1935–1955* (Evanston: Northwestern University Press, 1999), 68–69. Grams says that each episode in the mid-1940s cost between five and ten thousand dollars, an estimate that probably does not include purchase of network time; some actors were paid as much as five thousand dollars for an appearance. Martin Grams Jr., *The Official Guide to the History of the Cavalcade of America* (self-published, 1998), 35. See also Roland Marchand, *Creating the Corporate Soul: The Rise of Public Relations and Corporate Imagery in American Big Business* (Berkeley: University of California Press, 1998), 218–23.

8. Script for *Cavalcade of America*, episode no. 1, 1935, Cavalcade of America Collection, Hagley Museum and Library.

9. Script for *Cavalcade of America*, episode no. 9, 1935, Cavalcade of America Collection, Hagley Museum and Library.

10. Script for *Cavalcade of America*, episode no. 54, 1936, Cavalcade of America Collection, Hagley Museum and Library.

11. Ibid.

12. Script for *Cavalcade of America*, episode no. 117, 1938, Cavalcade of America Collection, Hagley Museum and Library.

13. Script for *Adventures in Research,* episode no. 185, United States National Museum, Department of Engineering and Industries, Records, 1891–1959, Record Unit 84, 15:49, Smithsonian Institution Archives.

14. Watson Davis served on an advisory board for this program; J. W. Studebaker to Watson Davis, April 14, 1936, Science Service Records, 173:6. See also Jeanette Sayre, *An Analysis of the Radiobroadcasting Activities of Federal Agencies,* Studies in the Control of Radio no. 3 (Cambridge: Harvard University, 1941), esp. 82.

15. Webster Prentiss True, *The Smithsonian Institution* (New York: Series Publishers, 1949), 332.

16. William D. Boutwell to C. G. Abbot, March 10, 1936, SI Secretary (RU46), 87:3.

17. J. W. Studebaker to C. G. Abbot, May 7, 1936, SI Secretary (RU46), 87:4. See also Webster True, Memorandum to the Secretary, May 23, 1936, SI Secretary (RU46), 15:7.

18. Sayre, *Analysis of Radiobroadcasting Activities,* 48–50.

19. "Prospectus for Complete Series," May 1936, SI Secretary (RU46), 15:7.

20. *The World Is Yours*, Listener Aid no. 5, 1939, United States National Museum, Department of Engineering and Industries, Records, 1891-1959, Record Unit 84, 15:48, Smithsonian Institution Archives.

21. C. P. Yare to the Smithsonian Institution, December 19, 1937, SI Secretary (RU46), 87:7.

22. Richard Philip Herget to Austin H. Clark, August 9, 1938, Clark Papers (NMNH). Secretary Abbot asked Austin Clark to supervise the project ("maintaining the accuracy and high standard of this series") while Webster True was absent during the summer of 1938. C. G. Abbot to Austin H. Clark, June 2, 1938, SI Secretary (RU46), 87:3.

23. C. G. Abbot to Lenox R. Lohr, July 31, 1936, SI Secretary (RU46), 15:6.

24. C. G. Abbot to Frank A. Taylor, March 30, 1939, United States National Museum, Department of Engineering and Industries, Records, 1891-1959, Record Unit 84, 15:47, Smithsonian Institution Archives.

25. Webster P. True to C. G. Abbot, May 20, 1939, SI Secretary (RU46), 15:7.

26. True, *Smithsonian*, 33.

27. Visitors to all four Smithsonian museums on the Mall totaled 1.9 million in 1936, 2.4 million in 1938, and 2.2 million in 1939. Data from Smithsonian Institution, U.S. National Museum, *Report on the Progress and Condition of the United States National Museum for the Year Ended June 30, 1936* (Washington, D.C.: U.S. Government Printing Office, 1937), and corresponding reports for 1937, 1938, and 1939 (published 1938, 1939, and 1940).

28. C. G. Abbot to Webster P. True, May 23, 1939, SI Secretary (RU46), 15:7.

29. Data from Sayre, *Analysis of Radiobroadcasting Activities*, 82.

30. C. G. Abbot to Harold Ickes, various correspondence in 1937, SI Secretary (RU46), 87:4; C. G. Abbot to NBC executives, 1937, SI Secretary (RU46), 15:6.

31. C. G. Abbot to D. B. Murray, July 10, 1941, SI Press, box 6.

32. H. B. Summers to W. P. True, January 23, 1942, SI Press, box 7.

33. William D. Boutwell to Thomas D. Rishworth, April 10, 1942, SI Press, box 7.

34. C. L. Menser to William D. Boutwell, April 14, 1942, SI Press, box 7.

35. Thornton W. Burgess to Austin H. Clark, August 14, 1935, Clark Papers (NMNH).

36. Austin H. Clark to Thornton W. Burgess, August 16, 1935, Clark Papers (NMNH).

37. Script for "The March of the Microbes" by Homer Calver, Science Service Records, 403:55.

38. F. R. Moulton, "Science by Radio," *Scientific Monthly* 47 (December 1938): 546.

39. Ibid., 547.

40. Ibid. See also Carroll Lane Fenton to Frank Thone, November 12, 1937, Science Service Records, 184:7.

41. James Stokley to Watson Davis, July 24, 1935, Science Service Records, 403:14.

42. Max Wylie, *Best Broadcasts of 1938-39* (New York: Whittlesey House, 1939), 302.

43. "Science Museum to Open Tuesday," *New York Times*, February 9, 1936; Waldemar Kaempffert, "Science: The Evolution of the Science Museum," *New York Times*, February 16, 1936.

44. A. William Bluem, *Documentary in American Television: Form, Function, Method* (New York: Hastings House, 1965), 62.

45. Bernard Victor Dryer, "Typhus," in *Radio Drama in Action: Twenty-Five Plays of a Changing World*, ed. Erik Barnouw (New York: Farrar and Rinehart, 1945), 323; Bluem, *Documentary in American Television*, 63.

46. Wylie, *Best Broadcasts of 1939-40*, 262-63.

47. " 'Unlimited Horizons,' a Weekly Broadcast," *Science* 92 (October 25, 1940): 374; publicity sheets for *Unlimited Horizons*, SI Press, box 17. See also Dunning, *On the Air*, 691; and Judith C. Waller, *Broadcasting in the Public Service* (Chicago: John C. Swift, 1943), 44.

48. Lyman Bryson, "Can We Put Science on the Air?," *Journal of Educational Sociology* 14 (February 1941): 368-69.

49. Script for January 9, 1942, broadcast of *Unlimited Horizons*, SI Press, box 17.

50. Raymond Fielding, *The American Newsreel, 1911-1967* (Norman: University of Oklahoma Press, 1972).

51. Author's estimate based on re-analysis of Leo Handel's data as cited in Fielding, *American Newsreel*, 3 and 290. Handel found that 1.1 percent of 1939 newsreels discussed science and 0.4 percent health. Leo A. Handel, *Hollywood Looks at Its Audience: A Report of Film Audience Research* (Urbana: University of Illinois Press, 1950).

52. Raymond Fielding, *The March of Time, 1935-1951* (New York: Oxford University Press, 1978), 10-13.

53. Bluem, *Documentary in American Television*, 36, 61; Fielding, *March of Time*. See also Thomas W. Hoffer, Robert Musburger, and Richard Alan Nelson, "Docudrama," in Brian G. Rose, ed., *TV Genres: A Handbook and Reference Guide* (New York: Greenwood Press, 1985), 185.

54. W. L. G. Joerg, *The Work of the Byrd Antarctic Expedition, 1928-1930* (New York: American Geographical Society, 1930), 2.

55. Ibid., 2.

56. See Jon D. Swartz and Robert C. Reinehr, *Handbook of Old-Time Radio: A Comprehensive Guide to Golden Age Radio Listening and Collecting* (Lanham, Md.: Scarecrow Press, 1993), 167; John Dunning, *Tune In Yesterday: The Ultimate Encyclopedia of Old-Time Radio, 1925-1976* (New York: Prentice-Hall, 1976), 5; and Harrison B. Summers, ed., *A Thirty-Year History of Programs Carried on National Radio Networks in the United States, 1926-1956* (New York: Arno Press and New York Times, 1971).

57. Benjamin C. Gruenberg, *Science and the Public Mind* (New York: McGraw-Hill, 1935), 1-2.

58. Quoted in Ray Barfield, *Listening to the Radio, 1920-1950* (New York: Praeger, 1996), 62.

59. C. D. Wagoner, General Electric Company News Bureau, "To Relatives and Friends of the Byrd Expedition," December 29, 1939, Science Service Records, 217:3. Watson's message on March 13, 1940, informed Malcolm that the emperor penguin collected for the National Zoo had arrived and was "safe and happy."

60. William Beebe, *Half Mile Down* (New York: Duell, Sloan and Pearce, 1951), 178. See also Robert Henry Welker, *Natural Man: The Life of William Beebe* (Bloomington: Indiana University Press, 1975), 129-31; and Carol Grant Gould, *The Remarkable Life of William Beebe: Explorer and Naturalist* (Washington, D.C.: Island Press/Shearwater Books, 2004), 303-16.

61. Beebe, *Half Mile Down*, 178-80.

62. Gould, *Remarkable Life*, 314.

63. Beebe, *Half Mile Down*, 176-80; Welker, *Natural Man*, 129-31; Robert J. Brown, *Manipulating the Ether: The Power of Broadcast Radio in Thirties America* (Jefferson, N.C.: McFarland, 1998), 136.

64. L. N. Diamond, "Interpreting Science to the Public," *Scientific Monthly* 40 (April 1935): 373.

65. Welker, *Natural Man*, 129.

66. "Eclipse Expedition to the South Seas," *Science* 85 (March 12, 1937): 257-58; "Broadcasts of the Eclipse," *Science* 85 (April 16, 1937): 377-78. See also "Wherever Big News Broke—NBC Was There," Science Service Records, 382:30.

67. *A Resume of CBS Broadcasting Activities During 1937* (New York: CBS, 1938).

68. National Broadcasting Company, "Year End Survey, No. 4—Special Events," December 30, 1938, Science Service Records, 382:25.

69. "Wherever Big News Broke—NBC Was There," Science Service Records, 382:30.

70. Brown, *Manipulating the Ether*, 237.

71. "'Doc' Jack Benny Presents Version of 'Yellow Jack,'" NBC Press Release, October 4, 1938, Science Service Records, 382:25.

72. Rosalynn D. Haynes, *From Faust to Strangelove: Representations of the Scientist in Western Literature* (Baltimore: Johns Hopkins University Press, 1994).

73. Watson Davis to H. E. Howe, August 4, 1936, Science Service Records, 175:6.

74. See Brown, *Manipulating the Ether*, chaps. 10–13.

75. Martha Strayer, "Realistic 'War' Broadcast Panics Radio Listeners," *Washington Daily News*, October 31, 1938.

76. An American Institute of Public Opinion poll conducted in December 1938 found that 26 percent of those who listened to *The War of the Worlds* believed that they were listening to a real news report. Hadley Cantril, *Public Opinion 1935–1946* (Princeton: Princeton University Press, 1951). See also Brown, *Manipulating the Ether*, 203, 225 and 238.

77. Hadley Cantril, *The Invasion from Mars: A Study in the Psychology of Panic* (Princeton: Princeton University Press, 1940), 70–71. CBS censors had required Welles to alter some but not all names of scientific institutions (e.g., changing "Museum of Natural History" to "Natural History Museum"). Brown, *Manipulating the Ether*, 205–8.

78. Hadley Cantril averaged the Gallup Poll estimate of nine million adults and the C. E. Hooper rating of four million; he then calculated the number who were "frightened" by extrapolating from his own interview sample. Cantril, *Invasion from Mars*, 47–56.

79. Cantril, *Invasion from Mars*.

80. Brown, *Manipulating the Ether*, 236.

81. Cantril, *Invasion from Mars*, 158.

82. Louis Reid, "Broadcasting Books: Drama, Fiction, and Poetry on the Air," *Saturday Review of Literature* 19 (January 14, 1939): 14.

83. Raymond Williams, *Television: Technology and Cultural Form* (New York: Schocken Books, 1975).

CHAPTER EIGHT

1. E. E. Slosson, "Memorandum to Mr. Davis," July 1, 1927, Science Service Records, 108:4.

2. E. E. Slosson, "Report to the Board of Trustees of Science Service," April 26, 1923, Merriam Papers, 194:1.

3. R. M. Langer to Watson Davis, April 24, 1934, Science Service Records, 1920s–1970s, Accession 90-105, 1:5, Smithsonian Institution Archives.

4. Charles Lewis Gazin was a curator in the Division of Vertebrate Paleontology until 1984; he took leave during World War II to serve in the Army Air Force performing intelligence work.

5. Watson Davis to Charles L. Gazin, July 19, 1934, Charles Lewis Gazin Papers, 22:19, Smithsonian Institution Archives.

6. Watson Davis to Margaret Cuthbert, December 4, 1934, Science Service Records, 167:11.

7. Watson Davis to J. N. Heiskell, October 22, 1936, Science Service Records, 424:14.

8. Margaret Cuthbert to Watson Davis, June 25, 1934, Science Service Records, 167:11.

9. Watson Davis to Margaret Cuthbert, October 15, 1934, Science Service Records, 167:11.

10. Margaret Cuthbert to Watson Davis, November 23, 1934, Science Service Records, 167:11.

11. Watson Davis to Margaret Cuthbert, November 27, 1934, Science Service Records, 167:11.

12. Margaret Cuthbert to Watson Davis, November 30, 1934, Science Service Records, 167:11.

13. Watson Davis to Malcolm M. Willey, September 17, 1936, Science Service Records, 181:4.

14. Jane Stafford to Watson Davis, June 15, 1934, Science Service Records, 154:1.

15. Watson Davis telegram to Robert Potter, September 12, 1934, Science Service Records, 158:1. See also Frank Thone's handwritten note on typescript of Davis telegram. Davis, Potter, and Jane Stafford all served as NASW presidents.

16. Robert D. Potter to Henry A. Barton, May 2, 1935, Science Service Records, 162:3; remarks by David Dietz, transcript of Symposium on Science and the Press held at AAAS Annual Meeting, Pittsburgh, 1934, Clark Papers (NMNH); Edmund S. Conklin to Willard C. Olson, January 3,

1938, and Howard W. Blakeslee to Edmund S. Conklin, January 10, 1938, American Psychological Association Records, box 35, folder "Committee on Publicity and Public Relations, 1933-1941," Manuscript Division, Library of Congress.

17. Around the same time, AAAS became involved in the popular news business, sponsoring *Science in the News*, a fifteen-minute science education talk, from 1936 to 1940. Michael M. Sokal, "Promoting Science in a New Century: The Middle Years of AAAS," in Sally Gregory Kohlstedt, Michael M. Sokal, and Bruce V. Lewenstein, *The Establishment of Science in America* (Brunswick: Rutgers University Press, 1999); Harrison B. Summers, ed., *A Thirty-Year History of Programs Carried on National Radio Networks in the United States, 1926-1956* (1958; New York: Arno Press and New York Times, 1971); Jon D. Swartz and Robert C. Reinehr, *Handbook of Old-Time Radio: A Comprehensive Guide to Golden Age Radio Listening and Collecting* (Metuchen, N.J.: Scarecrow Press, 1993); Frank Buxton and Bill Owen, *The Big Broadcast, 1920-1950* (Metuchen, N.J.: Scarecrow Press, 1997).

18. Watson Davis to Austin H. Clark, December 7, 1935, Science Service Records, 162:10.

19. John C. Duncan to Robert D. Potter, September 18, 1935, Science Service Records, 163:11.

20. Robert D. Potter to A. H. Joy, September 21, 1935, Science Service Records, 163:11.

21. Margaret Cuthbert to Watson Davis, Nov 30, 1934, Science Service Records, 167:11.

22. Watson Davis to George A. Pettit, August 6, 1934, Science Service Records, 157:14. By September 1934 broadcasts for the CBS series could be initiated from Boston, Los Angeles, San Francisco, Philadelphia, and Chicago, in addition to New York and Washington.

23. J. W. Foster to Thomas L. Sidlo, June 28, 1927, Science Service Records, 108:4.

24. E. E. Slosson, "Memorandum to Mr. Davis," July 1, 1927, Science Service Records, 108:4.

25. Frank Thone to George A. Pettitt, March 19, 1925, Science Service Records, 403:45.

26. Watson Davis to N. C. Nelson, September 11, 1935, Science Service Records, 403:12 and 13.

27. Henry J. Wing to Watson Davis, October 4, 1935, Science Service Records, 403:13.

28. Philip R. White to Emily C. Davis, January 18, 1938, Science Service Records, 204:3.

29. Broadcast of July 2, 1935. Watson Davis to Fred O. Tonney, June 19, 1935, Science Service Records, 403:21.

30. "Information Memorandum on Progress of Science Service January 15, 1936" (draft), Science Service Records, 4:2; Watson Davis to Edgar B. Howard, November 5, 1935, Science Service Records, 403:17.

31. Watson Davis to R. W. Wood, November 29, 1935, Science Service Records, 170:3.

32. Emily Cleveland Davis (b. 1898) was not related to Watson. She joined the Science Service staff in the 1920s, taking over the archeology/anthropology beat, and left during World War II to work for the U.S. Department of Agriculture. Davis published several books on archeology, including *Ancient Americans: The Archeological Story of Two Continents* (1931) and, with R. V. D. Magoffin, *Magic Spades: The Romance of Archeology* (1929).

33. When Murrow moved to London in spring 1937 to become CBS foreign news chief in London, Sioussat and her new boss, Sterling Fisher, took over supervision of public interest programs like the Science Service talks. Sioussat left that division in 1941 and eventually had her own CBS show called *Table Talk with Helen Sioussat*. See Marion Marzolf, *Up from the Footnote: A History of Women Journalists* (New York: Hastings House, 1977), 134-36; Donna L. Halper, *Invisible Stars: A Social History of Women in American Broadcasting* (Armonk, New York: M. E. Sharpe, 2001), 103; and Helen J. Sioussat Digital Gallery, "Taking a Leading Role," Library of American Broadcasting, University of Maryland, 2005, http://www.lib.umd.edu/LAB/exhibits/leadingrole.

34. "America 8,000 B.C.," script for December 3, 1935, broadcast of *Adventures in Science*, Science Service Records, 403:17.

35. Watson Davis to T. W. Church, December 6, 1936, Science Service Records, 167:2.

36. Final draft of "Pure Water" script for December 10, 1935, broadcast of *Adventures in Science*, Science Service Records, 403:18.

37. Jane Stafford to Thomas Parran, February 10, 1936, and Thomas Parran to Jane Stafford, February 18, 1936, Science Service Records, 177:12.

38. Scripts and correspondence in Science Service Records, 403:60.

39. Watson Davis to H. Bonnet, February 8, 1938, Science Service Records, 193:3.

40. Script for radio talk ("The Romance of Tung Oil"), Science Service Records, 194:6.

41. Watson Davis to G. J. Peirce, April 24, 1935, Science Service Records, 403:44. Saturday afternoon broadcasts in April had been preempted for baseball games.

42. Frank Thone to Anselm M. Keefe, May 18, 1935, Science Service Records, 403:43.

43. Remarks by Thomas R. Henry in "Edited Transcript of Symposium on Science and the Press held at AAAS Annual Meeting, Pittsburgh, 1934," 54, Clark Papers (NMNH).

44. Frank Thone to O. C. Durham, September 17, 1935, Science Service Records, 172:12.

45. Emily C. Davis to J. B. Kincer, June 8, 1936, Science Service Records, 403:61.

46. *A Resume of CBS Broadcasting Activities During 1937* (New York: CBS, 1938), 1.

47. Ibid., 5. See also Sterling Fisher, "The Radio and Public Opinion," *Public Opinion Quarterly* 2 (January 1938): 81.

48. Sterling Fisher to Watson Davis, June 1, 1937, and Watson Davis to Sterling Fisher, June 3, 1937, Science Service Records, 190:6; Watson Davis to Helen J. Sioussat, April 6, 1938, 194:10.

49. Lenox Lohr, "Some Social and Political Aspects of Broadcasting," address to U.S. Chamber of Commerce, May 4, 1938, 6, Science Service Records, 383:22.

50. Sterling Fisher, telegram to Watson Davis, April 8, 1938, Science Service Physics News Morgue Files, Accession 06-134, 1:9, Smithsonian Institution Archives.

51. Early version of script for August 12, 1938, broadcast of *Adventures in Science*, Science Service Records, 386:1.

52. Script for August 12, 1938, broadcast of *Adventures in Science*, Science Service Records, 386:1.

53. Script for August 19, 1938, broadcast of *Adventures in Science*, Science Service Records, 386:1.

54. A. M. Sperber, *Murrow: His Life and Times* (New York: Freundlich Books, 1986). See also Gerd Horten, *Radio Goes to War: The Cultural Politics of Propaganda during World War II* (Berkeley: University of California Press, 2002), 30–31.

55. Script for September 16, 1938, broadcast of *Adventures in Science*, Science Service Records, 386:1.

56. Sterling Fisher to Watson Davis, January 6, 1939, Science Service Records, 385:11.

57. Helen J. Sioussat to Watson Davis, January 20, 1939, Science Service Records, 385:22. Replying to her request, Davis wrote "I am now working on this program and will give you details as soon as they are worked out." Watson Davis to Helen J. Sioussat, January 31, 1939, Science Service Records, 385:22.

58. Memorandum to Watson Davis from Miss Nata Addis, May 25, 1939, Science Service Records, 385:25. See also Sterling Fisher, telegram to Watson Davis, August 11, 1939, and Watson Davis to Sterling Fisher, September 8, 1939, Science Service Records, 385:47.

59. Sterling Fisher, telegram to Watson Davis, March 8, 1939, Science Service Records, 385:15.

60. Watson Davis to Sterling Davis, March 11, 1939, Science Service Records, 385:15.

61. Draft outline of talk by Lewis W. Waters, n.d., Science Service Records, 385:15.

62. For example, Watson Davis to Brooks Darlington (DuPont advertising department), July 25, 1938, Science Service Records, 194:13; and James S. Little to Watson Davis (brokering an appearance by Willis Carrier), December 5, 1940, Science Service Records, 388:28.

63. William Wight to Watson Davis, March 2, 1940, Science Service Records, 386:24.

64. Helen J. Sioussat to Watson Davis, November 28, 1940, Science Service Records, 388:19. Sioussat suggested that Davis contact the public relations firm of Ivy Lee and T. J. Ross regarding a

potential appearance by Philco's chief engineer. Davis set up the January 23, 1941, program with E. O. Thompson, Philco's director of mechanical research.

65. Sterling Fisher, telegram to Watson Davis, January 24, 1939, Science Service Records, 385:9.

66. Script for November 13, 1939, broadcast of *Adventures in Science*, Science Service Records, 386:2.

67. See Watson Davis to W. Lee Lewis, September 26, 1939; Watson Davis, telegram to W. Lee Lewis, October 16, 1939; W. Lee Lewis to Watson Davis, October 16, 1939; and W. Lee Lewis to Watson Davis, October 25, 1939, Science Service Records, 386:2. In his October 16 letter Lewis explained that "at last a long life of chemicals has overtaken my vocal chords [*sic*] and I cannot rely on them any more." Davis implied on air that Lewis canceled because of last-minute "illness," and the CBS press release also downplayed the situation: "LARYNGITIS played an unusual role on CBS programs this week. When Dr. W. Lee Lewis, famous chemist, found no sound would come out of his larynx, he had his wife read his talk on poison gases on 'Adventures in Science.' " *CBS News and Notes* (November 16, 1939), Science Service Records, 386:2.

68. The program involved a switchover to WBBM-Chicago, where Mrs. Lewis was interviewed by the station announcer, reading from the script prepared by Davis.

69. A. F. Blakeslee to Watson Davis, December 9, 1939, Science Service Records, 386:17.

70. A. F. Blakeslee, telegram to Watson Davis, January 16, 1940, Science Service Records, 386:17. The discussion reused, at Blakeslee's suggestion, text from an essay he had written for the Associated Press in 1932.

71. John C. Burnham, *How Superstition Won and Science Lost: Popularizing Science and Health in the United States* (New Brunswick: Rutgers University Press, 1987), 43.

72. Watson Davis, "A Survey of the Interpretation of Science to the Public," 39, Science Service Records, 384:22; Paul F. Lazarsfeld, *Radio and the Printed Page* (1940; New York: Arno Press, 1971), 207-13.

73. Correspondence and scripts in Science Service Records, 380:57, 58, 59, and 60.

74. Script for November 20, 1939, broadcast of *Adventures in Science*, Science Service Records, 386:4.

75. Script for December 25, 1939, broadcast of *Adventures in Science*, Science Service Records, 386:10.

CHAPTER NINE

1. Remarks by Capt. J. W. Hellweg, U.S. Naval Observatory, "Science Service Conference. II," *Science* 76 (August 26, 1932): 183.

2. Watson Davis to William E. Ritter, November 21, 1941, Science Service Records, 272:9.

3. Harlow Shapley to Watson Davis, December 8, 1941, Science Service Records, 388:63.

4. Script for December 13, 1941, broadcast of *Adventures in Science*, Science Service Records, 388:63.

5. Ibid.

6. Ibid.

7. Script for July 4, 1942, broadcast of *Adventures in Science*, Science Service Records, 389:26.

8. Harold C. Relyea, "Shrouding the Endless Frontier—Scientific Communication and National Security: The Search for Balance," in *Striking a Balance: National Security and Scientific Freedom*, ed. Harold C. Relyea (Washington, D.C.: American Association for the Advancement of Science, 1985), 77.

9. "Artificial Radioactivity Produced," *Daily Mail Report*, January 31, 1934; "Making Atoms Smash Themselves," "Science Today" feature, mailed February 3, 1934; "First Chemical Proof of Transmutation Comes as Sequel to Artificial Radioactivity," *Daily Mail Report*, February 23,

1934; "Man-Made Radioactivity Achieved: Where Will It Lead? Science Asks," *Daily Wire Report*, February 27, 1934—all in Science Service Physics News Morgue Files, Accession 06-134, 1:7, Smithsonian Institution Archives.

10. "Atomic Bullet Streams Aimed at the Secrets of Matter's Constitution," *Daily Mail Report*, January 18, 1934; "Sodium Yields Gamma Rays by Artificial Means . . . ," *Daily Mail Report*, October 19, 1934; "The Neutron Ray" (February 5, 1934) and "An Eighty-five Ton Machine Gun" (February 9, 1936); scripts for 1934 broadcasts—all in Science Service Physics News Morgue Files, Accession 06-134, 1:5 and 1:7, Smithsonian Institution Archives.

11. Postscript note to editors (signed "W. D."), "Atomic Bullet Streams Aimed at the Secrets of Matter's Constitution," *Daily Mail Report*, January 18, 1934, 2, Science Service Physics News Morgue Files, Accession 06-134, 1:5, Smithsonian Institution Archives.

12. H. H. Goldsmith, "The Literature of Atomic Energy of the Past Decade," *Scientific Monthly* 68 (May 1949): 291-92. See also Lawrence Badash, Elizabeth Hodes, and Adolph Tiddens, "Nuclear Fission: Reaction to the Discovery in 1939," *Proceedings of the American Philosophical Society* 130 (June 1986): 196-231; and Lawrence Badash, *Scientists and the Development of Nuclear Weapons: From Fission to the Limited Test Ban Treaty, 1939-1963* (Atlantic Highlands: Humanities Press, 1995), 12-21.

13. "Brilliant Italian Physicist Wins Nobel Prize in Physics . . . ," Science Service News Bulletin, November 9, 1938, Science Service Records, 385:7; "Winner of Nobel Prize for Physics Here with Family," *New York Times*, January 3, 1939. Science Service writer Robert D. Potter interviewed Fermi at the Wardman Park Hotel to prepare for the February 2 broadcast. See Potter's notes in Science Service Records, 385:7.

14. Badash, Hodes, and Tiddens, "Nuclear Fission," 211.

15. Watson Davis, "Is World on Brink of Releasing Atomic Power? . . . ," Science Service news story, January 30, 1939, Science Service Records, 385:7.

16. Headlines from the (Washington) *Evening Star*, quoted in Badash, *Scientists and the Development of Nuclear Weapons*, 26. See also Badash, Hodes, and Tiddens, "Nuclear Fission," 212. Badash and his colleagues credit Davis and Potter with making the first connection in print "between fission and the possibility of explosives." See Watson Davis and Robert D. Potter, "Atomic Energy Released," *Science News Letter* 35 (February 11, 1939): 86-87. Potter wrote the Science Service news release, titled "Exploding Uranium Atoms May Set Free Neutrons That Will in Turn Explode Other Atoms, in 'Cascade' Effect," February 24, 1939, that was published in *Science News Letter*, March 11, 1939. See also discussion in Spencer R. Weart, *Scientists in Power* (Cambridge: Harvard University Press, 1979), 88-89.

17. Script for February 2, 1939, broadcast of *Adventures in Science*, Science Service Records, 385:7.

18. Ibid.

19. Harold C. Relyea, "Information, Secrecy, and Atomic Energy," *New York University Review of Law and Social Change* 10 (1980-1981): 267. See also Badash, *Scientists and the Development of Nuclear Weapons*, and Weart, *Scientists in Power*.

20. Badash, Hodes, and Tiddens, "Nuclear Fission," 215; Badash, *Scientists and the Development of Nuclear Weapons*, 29.

21. Sterling Fisher to Watson Davis, March 7, 1939, Science Service Records, 385:16.

22. See Watson Davis correspondence with Hale Sparks, April 1939, Science Service Records, 385:16.

23. Script for May 9, 1940, broadcast of *Adventures in Science*, Alexander Wetmore Papers, 233:6, Smithsonian Institution Archives.

24. Spencer R. Weart, "Scientists with a Secret," *Physics Today* 29 (February 1976): 23-30; Spencer R. Weart and Gertrud Weiss Szilard, *Leo Szilard: His Version of the Facts, Selected Recollections and Correspondence* (Cambridge: MIT Press, 1978), 118-35; Weart, *Scientists in Power*, 146.

25. National Research Council, "Publication of Scientific Work under Emergency Conditions," July 1940, Science Service Records, 261:5.

26. Michael M. Sokal, "From the Archives," *Science, Technology & Human Values* 10 (Spring 1985): 24; Michael M. Sokal, "Restrictions on Scientific Publication," *Science* 215 (March 5, 1982): 1182. See also "Science Hush-Hushed," *Time*, May 11, 1942, 90; H. H. Goldsmith, "Literature of Atomic Energy," 291–98; and James Stacey Thompson, *The Technical Book Publisher in Wartimes* (New York: New York Public Library, 1942).

27. Watson Davis to Robert B. Jacobs, November 29, 1940, and Robert B. Jacobs to Watson Davis, December 4, 1940, Science Service Records, 261:5.

28. Watson Davis to Robert B. Jacobs, November 29, 1940, Science Service Records, 261:5.

29. Byron Price, "The American Way," in *Journalism in Wartime: The University of Missouri's Thirty-Fourth Annual "Journalism Week" in Print,* ed. Frank Luther Mott (1943; New York: Greenwood Press, 1984), 23–33.

30. Robert E. Summers, ed., *Wartime Censorship of Press and Radio* (New York: H. W. Wilson, 1942), 115; Patrick S. Washburn, "The Office of Censorship's Attempt to Control Press Coverage of the Atomic Bomb During World War II," *Journalism Monographs*, no. 120 (April 1990), esp. 4; Michael S. Sweeney, *Secrets of Victory: The Office of Censorship and the American Press and Radio in World War II* (Chapel Hill: University of North Carolina Press, 2001), 195–206. See also Paul S. Boyer, *Purity in Print: Book Censorship in America from the Gilded Age to the Computer Age,* 2nd ed. (University of Wisconsin Press, 2002), 271–72; and Mott, *Journalism in Wartime.*

31. Washburn, "Office of Censorship's Attempt"; Reginald R. Hawkins, *Technical Books and the War: An Exhibition* (New York: New York Public Library, 1943). See also James L. Baughman, *The Republic of Mass Culture: Journalism, Filmmaking, and Broadcasting in America Since 1941* (Baltimore: Johns Hopkins University Press, 1992), 1–9.

32. Washburn, "Office of Censorship's Attempt," esp. 33.

33. Watson Davis to Paul Bergen, December 30, 1942, Science Service Records, 232:1.

34. Washburn, "Office of Censorship's Attempt."

35. Having worked during the late 1960s as an assistant editor on two commercial technical magazines, I am keenly aware that editors strive to balance these interests in a responsible and ethical manner.

36. George D. Welles Jr. to S. D. Kirkpatrick, December 29, 1941, Science Service Records, 237:4.

37. Ibid.

38. S. D. Kirkpatrick to George D. Welles Jr., January 5, 1942; S. D. Kirkpatrick to George D. Welles Jr., January 14, 1942; and S. D. Kirkpatrick to Watson Davis, January 5, 1942, Science Service Records, 237:4.

39. S. D. Kirkpatrick to George D. Welles Jr., January 5, 1942, Science Service Records, 237:4.

40. See, for example, Vince Kiernan, *Embargoed Science* (Urbana: University of Illinois Press, 2006).

41. "Use of DDT as a Mosquito Larvicide" was originally scheduled to be published on page 2010 of *Journal of the American Medical Association,* April 22, 1944. See materials in Science Service Records, 267:2–3.

42. Watson Davis to Luther P. Eisenhart, August 12, 1942, Science Service Records, 234:6.

43. "Science Hush-Hushed," *Time*, May 11, 1942, 90.

44. Selman Waksman to Watson Davis, September 8, 1942 [letter 1], Science Service Records, 389:41.

45. Selman Waksman to Watson Davis, September 8, 1942 [letter 2], Science Service Records, 389:41.

46. Watson Davis to Selman Waksman, September 15, 1942, Science Service Records, 389:41.

47. "Sudden Burst of Radar Advertising," *Printers' Ink,* June 1943, 37–38.

48. "NOTE TO EDITORS AND BROADCASTERS" (S-2325), U.S. Office of Censorship, July 29, 1943. When Watson Davis circulated this bulletin to Science Service staff, one person added at the bottom "FYI—Army has stopped splurge of advertising" (Science Service Records, 272:5).

49. Washburn, "Office of Censorship's Attempt," 20–22, reprints these Superman drawings.

50. Louis N. Ridenour, "Military Secrecy and the Atomic Bomb," *Fortune* 32 (November 1945), 170–71; Washburn, "Office of Censorship's Attempt," 20–22. See also Paul Brians, *Nuclear Holocausts: Atomic War in Fiction, 1895–1984* (Kent, Ohio: Kent State University Press, 1987).

51. Harlow Shapley to Lyman Bryson, November 24, 1943, and Lyman Bryson to Harlow Shapley, December 2, 1943, Science Service Records, 245:6. That suspension lasted until June 1944, long past football season; broadcasts were again suspended from September 9 through December 9, 1944, and similar three-month interruptions continued for many years thereafter.

52. As quoted in Watson Davis to Harlow Shapley, September 7, 1943, Science Service Records, 264:4.

53. Harlow Shapley to Watson Davis, July 20, 1945, Science Service Records, 284:9.

54. Frank Thone to Harlow Shapley, July 22, 1945, Science Service Records, 284:9. Davis was traveling as part of a U.S. Department of State program to arrange Spanish and Portuguese translations of scientific and medical books. See Watson Davis, "Translated Books for and from Latin America," *Publishers' Weekly*, April 14, 1945.

55. Frank Thone to Watson Davis (in Mexico City), September 2, 1945, Science Service Records, 267:4.

56. Script for August 11, 1945, broadcast of *Adventures in Science*, final version, p. 2, Science Service Records, 391:27.

57. Frank Thone to J. E. Haynes, December 10, 1945, Science Service Records, 269:1.

58. *Serving through Science, a Series of Talks Delivered by American Scientists on the New York Philharmonic-Symphony Program* (New York: U.S. Rubber Company, 1946), v. See also Warren Weaver, ed., *The Scientists Speak* (New York: Boni & Gaer, 1947); and Columbia Broadcasting System, *Crescendo: A Chronicle of Radio's Unique Power to Move People to Direct Action* (New York: CBS, 1947).

59. *Serving through Science*, vi.

60. Watson Davis to Harlow Shapley, January 25, 1945, Science Service Records, 284:9.

61. For analysis of five thousand of those letters, see Columbia Broadcasting System, *Crescendo*.

62. *Serving through Science*, vii.

63. Raymond B. Fosdick, "A Layman Looks at Science," broadcast for November 4, 1945, reprinted in *Serving through Science*, 98.

64. Theodore Frederick Koop, *Weapon of Silence* (Chicago: University of Chicago Press, 1946), 285.

65. Harold P. Green, "Information Control under the Atomic Energy Act," in Relyea, *Striking a Balance*, 58.

66. See "American Society of Newspaper Editors Reports on Atomic Information Problems," *Bulletin of the Atomic Scientists* 4 (July 1948): 211–12; James J. Butler, "ASNE, Atomic Board Join in Security Study," *Editor & Publisher* 80 (December 6, 1947): 10; and Robert Edward Summers, ed., *Federal Information Controls in Peacetime* (New York: H. W. Wilson, 1949).

67. W. A. Shurcliff, *Bombs at Bikini: The Official Report of Operation Crossroads* (New York: William H. Wise, 1947).

68. Robert Lewis Shayon, *Operation Crossroads as Broadcast from the Coolidge Auditorium of the Library of Congress* (Washington, D.C.: Library of Congress, 1946).

69. Paul Boyer, *By the Bomb's Early Light: American Thought and Culture at the Dawn of the Atomic Age* (New York: Pantheon, 1985). In 1946 John Hersey's *Hiroshima* was serialized in the *New Yorker*, condensed by *Reader's Digest*, and offered as a Book-of-the-Month Club selection, and portions of the text were reprinted in newspapers and read on the radio. See also Weart, *Nuclear Fear*, and Winkler, *Life under a Cloud*.

70. On the fourth estate, see Douglass Cater, *The Fourth Branch of Government* (Boston: Houghton Mifflin, 1959). On the fifth estate, see Arthur D. Little, "The Fifth Estate," *Science* 60 (October 3, 1924): 299–306; Arthur D. Little, "The Fifth Estate," *Atlantic Monthly* (December 1924): 771–81; Don K. Price, *The Scientific Estate* (Cambridge: Harvard University Press/Belknap Press, 1965).

71. Walter J. Murphy to Watson Davis, March 29, 1946, Science Service Records, 281:4.

72. Daniel Wilkes to Watson Davis, August 3, 1946, Science Service Records, 286:7.

73. Frank Thone to James W. Brown, July 14, 1946, Science Service Records, 278:2.

74. Ibid.

75. Frank Thone, Science Service Wire Report for June 24, 1946, Science Service Records 1920s–1970s, Accession 90-105, box 39, folder "QC Atomic Tests"; Frank Thone, *Science News Letter*, July 6, 1946. This episode might give the impression that scientists like Thone were wary of DDT, but his correspondence indicates quite the opposite. In one letter, he recommended using an aerosol "DDT bomb" in kitchens and dining rooms to control insects like fruit flies, as long as "you don't spray it directly on food." Frank Thone to Mrs. M. F. Blank, November 6, 1946, Science Service Records, 276:3.

76. Norris W. Rakestraw, editor of *Journal of Chemical Education*, to Watson Davis, July 2, 1946, and Watson Davis to Norris W. Rakestraw, July 6, 1946, Science Service Records, 280:2.

77. The University of Chicago Round Table, *America and the Atomic Age, Special Twentieth Anniversary Pamphlet, February 1931 to February 1951* (Chicago: University of Chicago, 1951), 3–9.

78. The Federation of American Scientists worked with the Advertising Council to create public service messages and with local stations to produce documentaries. Alice Kimball Smith, *A Peril and a Hope: The Scientists' Movement in America, 1945–47*, rev. ed. (Cambridge: MIT Press, 1970), 282 and 296. See also Boyer, *By the Bomb's Early Light*, 68 and 76.

79. Watson Davis, handwritten notes, in Science Service Records, 391:33.

80. Script for December 1, 1945, broadcast of *Adventures in Science*, Science Service Records, 391:33.

81. Quoted in Judith C. Waller, *Radio: The Fifth Estate*, 2nd ed. (Boston: Houghton Mifflin, 1950), 213. Many historians credit documentaries like *The Sunny Side of the Atom* and *Atom and You* with shaping Americans' response to nuclear power. See Paul S. Boyer, *Fallout: A Historian Reflects on America's Half-Century Encounter with Nuclear Weapons* (Columbus: Ohio State University Press, 1998), 30; and Spencer R. Weart, *Nuclear Fear: A History of Images* (Cambridge: Harvard University Press, 1988).

82. Described in Boyer, *By the Bomb's Early Light*, 65.

83. Script for July 25, 1946, broadcast of *You and the Atom*, p. 1, Science Service Records, 392:25.

84. *CBS Network News Release*, "CBS Documentary Unit Program to Reveal Progress toward Better World . . . ," June 18, 1947, Science Service Physics News Morgue Files, Accession 06-134, 1:1, Smithsonian Institution Archives. See also Boyer, *By the Bomb's Early Light*, 299–301.

85. Irving J. Gitlin, "Radio and Atomic Energy Education," *Journal of Educational Sociology* 22 (January 1949): 329. See also Boyer, *By the Bomb's Early Light*, 299–300.

86. A. William Bluem, *Documentary in American Television: Form, Function, Method* (New York: Hastings House, 1965), 70–71; John Dunning, *Tune In Yesterday: The Ultimate Encyclopedia of Old-Time Radio, 1925–1976* (New York: Prentice-Hall, 1976), 495; John Dunning, *On the Air: The Encyclopedia of Old-Time Radio* (New York: Oxford University Press, 1998), 558; Thomas W. Hoffer, Robert Musburger, and Richard Alan Nelson, "Docudrama," in *TV Genres: A Handbook and Reference Guide*, ed. Brian G. Rose (New York: Greenwood Press, 1985), 186–87.

87. "Public Reaction to the Atomic Bomb and World Affairs, Report of the SSRC Subcommittee on Public Reactions to the Atomic Bomb and International Relations" (Cornell University, 1947), cited in Gitlin, "Radio and Atomic Energy Education," 328.

88. Gitlin, "Radio and Atomic Energy Education," 327–28.

89. Script for December 16, 1950, broadcast of *Adventures in Science*, p. 1, Science Service Records, 397:12. See also scripts for the July 15 and July 22, 1950, broadcasts of *Adventures in Science* ("Our Atomic Future," parts 1 and 2), Science Service Records, 397:20.

90. Script for April 29, 1950, broadcast of *Adventures in Science*, Science Service Records, 397:37.

91. Ernest L. Stebbins, "The Responsibility of the Radio Industry in Public Health," in New York Academy of Medicine, *Radio in Health Education*, 69-73. See also Columbia University, Bureau of Applied Social Research, *The People Look at Radio* (Chapel Hill: University of North Carolina Press, 1946), 56-48; and Paul F. Lazarsfeld and Patricia L. Kendall, *Radio Listening in America: The People Look at Radio—Again* (New York: Prentice-Hall, 1948).

92. Frank Thone, "Science in the Post-War World," April 1943, Science Service Records, 251:2.

93. The series aired first on MBS with commercial sponsorship (1945-1947) and then on ABC for one season as a sustaining program. Dunning, *On the Air*, 238.

94. Script for May 28, 1946, broadcast of *Frontiers of Science*, describing a paleontologist's identification of bones found during a building renovation. Science Service Records, 396:2 and 397:1.

95. See William L. Laurence, "Man Seen Ruling His Own Evolution," *New York Times*, January 4, 1947; and William L. Laurence, "Suggests Rodents May Inherit Earth," *New York Times*, January 5, 1947.

96. J. B. S. Haldane, handwritten notes for delivery, and retyped remarks inserted to script for January 4, 1947, broadcast of *Adventures in Science*, Science Service Records, 393:1.

CHAPTER TEN

1. "The Worm Turns," *The Johns Hopkins Science Review*, first broadcast July 21, 1952, JHSR Videotapes.

2. Lynn Poole, *Science via Television* (Baltimore: Johns Hopkins Press, 1950), 20 and 78.

3. Michael Tracey, "Non-Fiction Television," in *Television: An International History*, ed. Anthony Smith (Oxford: Oxford University Press, 1995), 121.

4. John C. Burnham, *How Superstition Won and Science Lost: Popularizing Science and Health in the United States* (New Brunswick: Rutgers University Press, 1987).

5. Data in this paragraph comes from Robert G. Finney, "Television," in *Handbook on Mass Media in the United States*, ed. Erwin K. Thomas and Brown H. Carpenter (New York: Greenwood Press, 1994), 169; Garth Jowett, *Film, The Democratic Art* (Boston: Focal Press, 1985), 348; and Sydney W. Head, *Broadcasting in America: Survey of Television and Radio* (Cambridge: Riverside Press, 1956), 158.

6. Larry James Gianakos, *Television Drama Series Programming: A Comprehensive Chronicle, 1947-59* (Lanham, Md.: Scarecrow Press, 1980), 3.

7. Head, *Broadcasting in America*, 158. See also James von Schilling, *The Magic Window: American Television, 1939-1953* (New York: Haworth Press, 2003).

8. Susan J. Douglas, *Listening In: Radio and the American Imagination* (New York: Times Books, 1999), 220-21.

9. Judy Dupuy, *Television Show Business* (Schenectady: General Electric, 1945), 68. In 1943, a series produced by Dupuy took local television viewers "backstage" to observe production of an *Excursions in Science* radio broadcast.

10. Harry Matthei, "Inventing the Commercial: The Imperium of Modern Television Advertising Was Born in Desperate Improvisation," *American Heritage* 48 (May/June 1997), text quoted from online version at http://americanheritage.com.

11. Tim Brooks and Earle Marsh, *The Complete Directory to Prime Time Network and Cable TV Shows, 1946-Present*, 6th ed. (New York: Ballantine Books, 1995), ix, 919. Some reference books

say the host was Guthrie McClintock. Brooks and Marsh identify him as "Dr. Miller McClintock." U.S. Rubber sponsored a similar educational show locally in New York in 1945. *Serving through Science* was also the title of a book of transcripts from U.S. Rubber's 1945-48 radio talk series.

12. Judith C. Waller, *Radio: The Fifth Estate* (Boston: Houghton Mifflin, 1946), 171.

13. Lyman Bryson, "The Responsibility for Program Planning," in New York Academy of Medicine, *Radio in Health Education* (New York: Columbia University Press, 1945), 85.

14. Roy K. Marshall, "Televising Science," *Physics Today* 2 (January 1949): 26.

15. Leo Geier makes this point in *Ten Years with Television at Johns Hopkins* (Baltimore: Johns Hopkins University, 1958). See also the criteria for media visibility described in Rae Goodell, *The Visible Scientists* (Boston: Little Brown, 1977).

16. Brooks and Marsh, *Complete Directory*, 732-33. In 1949, Marshall became director of Morehead Planetarium in North Carolina.

17. Poole, *Science via Television*, 2-3.

18. Marshall, "Televising Science," 28. See also Roy K. Marshall, *The Nature of Things* (New York: Henry Holt, 1951).

19. Ibid., 26.

20. WMAR-TV was owned by Mencken's newspaper, the *Sun*. Robert B. Cochrane to Watson Davis, October 8, 1947, Science Service Records, 287:6.

21. See Jack Gould, "Science Offered over Television," *New York Times*, October 18, 1950; Leonard D. Pigott, "Biology Reaches a New Horizon: Science on Television," *A.I.B.S. Bulletin* 1 (July 1951): 7; "Lynn Poole, Won Early TV Prizes," *New York Times*, April 16, 1969; Sue De Pasquale, "Live from Baltimore—It's the Johns Hopkins Science Review!," *Johns Hopkins Magazine*, February 1995; Poole, *Science via Television*; and Geier, *Ten Years with Television*.

22. See correspondence between Poole and provost P. Stewart Macauley in JHU News Office, 1:1.

23. Robert M. Yoder, "TV's Shoestring Surprise," *Saturday Evening Post*, August 21, 1954, 30, 90-92.

24. Irving Stettel, ed., *Top TV Shows of the Year, 1954-1955* (New York: Hastings House, 1955), 7.

25. Lynn Poole, as quoted in R. D. Heldenfels, *Television's Greatest Year—1954* (New York: Continuum, 1994), 182.

26. Lynn Poole, "Science on Video," *New York Times*, March 9, 1952.

27. Poole's introductory remarks on "Toys and Science," broadcast of January 20, 1954, JHSR Videotapes. Script reprinted in Stettel, *Top TV Shows of the Year*, 8.

28. "Conquest of Pain—Anesthesiology," broadcast of September 29, 1954, JHSR Videotapes. Script reprinted in Stettel, *Top TV Shows of the Year*, 221.

29. Poole, *Science via Television*, 25.

30. Fellowship brochure, JHU News Office, Record Group 10.020, Series 10, ("Lynn Poole's Scrapbook 1952-1953"), 1:5.

31. Reprint of article "From Mousetraps to the Moon with Johns Hopkins File 7," n.d. (between 1956 and 1960), JHU News Office, Record Group 10.020, Series 10 ("Lynn Poole/Television"), 1:3.

32. Geier, *Ten Years with Television*, 17.

33. Yoder, "TV's Shoestring Surprise," 30, 90; Stettel, *Top TV Shows of the Year*, 6.

34. Author's analysis of "Guests on Johns Hopkins Television Series, 1948-60," n.d., JHU News Office, Record Group 10.020, Series 10 ("Lynn Poole/Television"), 1:1. The list contains the names of 459 individuals who appeared on 425 of the 491 total broadcasts in the four Hopkins series; of these, 32 were women, but only 11 were scientists or women engaged in some science-related activity or profession (other occupations included museum art restorer, dog trainer, television executive, BBC announcer, and stuntwoman). There is no reason to suspect that the 66 broadcasts omitted from the news office list featured a greater proportion of women scientists.

35. Chemist Maria Telkes appeared on *The Johns Hopkins Science Review* on June 3, 1954 ("Heating Houses with the Sun's Rays") as one of two guests; on March 26, 1955, she was the sole guest interviewed on *Adventures in Science*.

36. Television's premier science series, *NOVA*, demonstrates the persistence of marginalizing stereotypes. During its first twenty-eight seasons, *NOVA* presented thirty-two new "biography" programs (6 percent of all nonrepeat programs). Only two of those programs focused exclusively on the lives and careers of women scientists. Author's analysis of themes of 503 new (not repeat) *NOVA* programs from classifications posted on the *NOVA* Web site. See also Jocelyn Steinke and Marilee Long, "A Lab of Her Own? Portrayals of Female Characters on Children's Educational Science Programs," *Science Communication* 18 (December 1996): 91-115.

37. Gary Schulz, "Is Video a Good Education Medium?" *Wisconsin Alumnus* 53 (June 1952): 8-10.

38. Ibid., 8.

39. Heldenfels, *Television's Greatest Year*, 179. For analysis of how commercial television determined the fate of educational broadcasting, see James L. Baughman, *Same Time, Same Station: Creating American Television, 1948-1961* (Baltimore: Johns Hopkins University Press, 2007).

40. Schulz, "Is Video a Good Education Medium?" 8.

41. Poole, *Science via Television*, 3-4.

42. Benjamin Draper, "Producing 'Science in Action,' " in Benjamin Draper, ed., *The 'Science in Action' TV Library*, vol. 1 (New York: Merlin Press, 1956), x; Lawrence E. Davies, " 'Science in Action': Western Show Entertains and Gives Information," *New York Times*, July 5, 1953.

43. Ibid., xi.

44. Ibid., xv.

45. Draper, "Producing 'Science in Action'," xv.

46. Poole, *Science via Television*, 79.

47. Gregg Mitman, *Reel Nature: America's Romance with Wildlife on Film* (Cambridge: Harvard University Press, 1999), 174.

48. Ibid., 144.

49. Robert M. Weitman, "Television Is Coming of Age," in Stettel, *Top TV Shows of the Year*, xv.

50. National Association of Educational Broadcasters (NAEB), *Lincoln Lodge Seminar on Educational Television: Proceedings* (Urbana, Illinois: NAEB, 1953), 3.

51. Triangle Publications, *The University of the Air: Commercial Television's Pioneer Effort in Education* (Philadelphia: Triangle Publications, 1959).

52. Brooks and Marsh, *Complete Directory*, 1122; McNeil, *Total Television*, 905; "What in the World: The Original TV Show," University of Pennsylvania Museum of Archeology and Anthropology Web site, http://www.upenn.edu/museum/Games/whatintheworld.html, accessed May 4, 2007.

53. Philip Hamburger, "Experts," *New Yorker* (May 3, 1952); Gary Schulz, "Is Video a Good Education Medium?"9.

54. William Melody and Wendy Ehrlich, "The History of Children's Television," in William Melody, *Children's Television: The Economics of Exploitation* (New Haven: Yale University Press, 1973), 361.

55. Ibid.

56. Ibid.

57. *Kaleidoscope* is distinguished primarily for being one of the first series of any type to be placed in syndication. Hal Erikson, *Syndicated Television: The First Forty Years, 1947-1987* (Jefferson, N.C.: McFarland, 1989), 75. Another syndicated series, *Junior Science*, hosted by George Wendt, focused on creativity, discovery, and experimentation.

58. *Watch Mr. Wizard* premiered live in 1951 from WNBQ-TV in Chicago, and then ran on NBC until 1965. After that series was canceled, Herbert teamed with General Motors Research

Laboratories and the U.S. National Science Foundation (NSF) to produce over two hundred eighty-second science "inserts" for local news shows; *How About...* was syndicated to 135 stations in 1981. Next, Herbert created *Mister Wizard's World* for the Nickelodeon channel and, in the 1990s, collaborated with NSF and Nickelodeon on a fifteen-minute series for science teachers. George Alexander, "Don Herbert: Gee Wiz!," *SciQuest*, May/June 1981, 21-28; Cleveland Amory, "Mr. Wizard," *TV Guide*, April 22, 1972, 48; Brooks and Marsh, *Complete Directory*, 1111; Jeffrey Davis, *Children's Television, 1947-1990* (Jefferson, N.C.: McFarland, 1995), 151-53; Diane Dismuke, "Don 'Mr. Wizard' Herbert: Still a Science Trailblazer," *NEA Today* 12 (April 1994): 9; Alex McNeil, *Total Television*, 4th ed. (New York: Penguin Books, 1996), 562; Vincent Terrace, *The Complete Encyclopedia of Television Programs 1947-1976*, 2 vols. (South Brunswick and New York: A. S. Barnes, 1976).

59. Dismuke, "Don 'Mr. Wizard' Herbert," 9.

60. Herbert concluded each commercial by saying, "As you know, at General Electric, progress is our most important product." "How a Television Commercial Is Made," *General Electric Review* (September 1956), reprinted in *Mass Communications*, 2nd ed., ed. Wilbur Schramm (Urbana: University of Illinois Press, 1960), 161-74.

61. From the 1952 Peabody Awards Digest, quoted in Library of Congress catalog description of *Omnibus*.

62. Neal Gabler, *Walt Disney: The Triumph of the American Imagination* (New York: Knopf, 2006), 504-12; Michael R. Real, *Mass-Mediated Culture* (New York: Prentice-Hall, 1977); Steven Watts, *The Magic Kingdom: Walt Disney and the American Way of Life* (Boston: Houghton Mifflin, 1997). Until Disney established its own cable channel, the programs ran in primetime on broadcast networks—as *Disneyland* and *Walt Disney Presents*, (ABC, 1954-1961), as *Walt Disney's Wonderful World of Color*, *The Wonderful World of Disney*, and *Disney's Wonderful World* (NBC, 1961-1981), *Disney's Wonderful World* (CBS, 1981-1983), ABC as *The Disney Sunday Movie*, (1986-1988), and *The Magical World of Disney* (NBC, 1988-1990).

63. Elizabeth Walker Mechling and Jay Mechling, "The Atom According to Disney," *Quarterly Journal of Speech* 81 (November 1995): 437.

64. Author's analysis of nature, space, and other science-related segments on *Disneyland* television series, seasons 1 (1954-1955) through 16 (1969-1970), based on program descriptions in Larry James Gianakos, *Television Drama Series Programming: A Comprehensive Chronicle, 1947-59* (Metuchen, N.J.: Scarecrow Press, 1980), and Larry James Gianakos, *Television Drama Series Programming: A Comprehensive Chronicle, 1959-75* (Metuchen, N.J.: Scarecrow Press, 1978).

65. Watts, *Magic Kingdom*, 309.

66. Gabler, *Walt Disney*, 517. See also Watts, *Magic Kingdom*.

67. Gabler, *Walt Disney*, 517; Watts, *Magic Kingdom*, 309 and 311.

68. Watts, *Magic Kingdom*, 312; Haber, *Our Friend the Atom*, 13. See also Allan M. Winkler, *Life under a Cloud: American Anxiety about the Atom* (New York: Oxford University Press, 1993), 140-41.

69. Mechling and Mechling, "Atom According to Disney"; Watts, *Magic Kingdom*, 312, 372. Disney established a science department headed by science writer Heinz Haber. *Our Friend the Atom* premiered on television on January 23, 1957. Within a few months, Heinz Haber's book version (New York: Simon & Schuster, 1957) sold two hundred thousand copies. See Spencer R. Weart, *Nuclear Fear: A History of Images* (Cambridge: Harvard University Press, 1988), 169 n. 42.

70. "Special" is the industry term for a specially produced program that preempts regular programming. Most are single documentaries or dramas, but may be produced and presented with a unifying identifier, as with the *Bell Telephone Science Series*.

71. Gilbert found that discussion began around 1951, encouraged by AT&T board member Vannevar Bush. James Burkhart Gilbert, *Redeeming Culture: American Religion in an Age of Science* (Chicago: University of Chicago Press, 1997), 202.

72. Bell also subsidized *The Restless Sea*, produced by Disney Studios, but with a new host and narrator. *The Restless Sea* premiered in 1964 with respectable ratings but was not included in this

analysis because it was not distributed widely, was not in the video set marketed to educators, and was available for sale only in an edited, thirty-four-minute version.

73. Phrases from opening and closing credits.

74. Data cited in Gilbert, *Redeeming Culture*, 223, 367 n. 62.

75. Ibid., 206-8.

76. Ibid., 212.

77. Ibid., 213.

78. After the success of *Shakespeare on TV* (1954) and *Now and Then* (1954-1955), Baxter declined many lucrative offers to perform on commercial television before agreeing to the Bell project. Sydney Head, *Broadcasting in America: Survey of Television and Radio* (Cambridge: Riverside Press, 1956), 410 n. 12.

79. David Templeton, "Weird Science," *Sonoma County Independent*, September 23-29, 1999.

80. Gilbert, *Redeeming Culture*.

81. The board suggested that Dr. Research say: "Physically we believe you descended from among an evolutionary line of lesser animals but at some point near the end of this evolution a miracle occurred and you became endowed with the human spirit which distinguishes you from all animals." Gilbert, *Redeeming Culture*, 214.

82. Ibid., 200, 217.

83. R. D. Heldenfels, *Television's Greatest Year—1954* (New York: Continuum, 1994), 177.

84. Lawrence W. Lichty, "Success Story," *Wilson Quarterly* 5 (Winter 1981): 63-64.

85. Head, Broadcasting in America, 408.

86. Lichty, "Success Story," 62.

87. Thomas Streeter, *Selling the Air: A Critique of the Policy of Commercial Broadcasting in the United States* (Chicago: University of Chicago Press, 1996), 277.

88. Ibid., 276.

89. Edward R. Murrow, as quoted in Sander Vanocur, "TV's Failed Promise," *Center Magazine*, November/December 1971, 46.

EPILOGUE

1. Lyman Bryson, "The Responsibility for Program Planning," in New York Academy of Medicine, *Radio in Health Education* (New York: Columbia University Press, 1945), 83.

2. Factors such as loosening of government controls also played a role. See Elizabeth L. Eisenstein, *The Printing Press as an Agent of Change*, 2 vols. (Cambridge: Cambridge University Press, 1979).

3. Austin H. Clark to James McKeen Cattell, December 2, 1925, Clark Papers (SIA), 3:6.

4. Paul F. Lazarsfeld, "The Effects of Radio on Public Opinion," in Douglas Waples, ed., *Print, Radio, and Film in a Democracy* (Chicago: University of Chicago Press, 1942), 66.

5. Ibid., 66-69.

6. Sydney W. Head, *Broadcasting in America: Survey of Television and Radio* (Cambridge: Riverside Press, 1956), 404.

7. Watson Davis, Speech to Institute on Collecting Science Literature for General Reading, University of Illinois, November 7, 1960, p. 4, Science Service Records, 368:2.

8. David Buckingham, *Children Talking Television: The Making of Television Literacy* (London: Falmer Press, 1993), 18 and 39.

9. Milton Chen, "Television and Informal Science Education: Assessing the Past, Present, and Future of Research," in Valerie Crane, ed., *Informal Science Learning: What the Research Says about Television, Science Museums, and Community-Based Projects* (Watertown: Research Communications, 1994), 18.

10. David T. Suzuki, "Information Overload: More TV Science Could Add to the Confusion," *SIPIscope* 14 (January-February 1986): 4.

11. V. O. Key Jr., *Public Opinion and American Democracy* (New York: Alfred A. Knopf, 1961), 366–67.

12. Tony Schwartz, *The Responsive Chord* (New York: Anchor Books, 1973), 17–18 and 146.

13. Frank Thone, "The Press as an Agency for the Diffusion of Science," text of speech to the American Association for Adult Education, May 21, 1936, Science Service Records, 4:2.

Bibliography

MANUSCRIPT COLLECTIONS

Smithsonian Institution Archives

Alexander Wetmore Papers, circa 1848–1979 and undated, Record Unit 7006.
Austin Hobart Clark Papers, 1883–1954 and undated, Record Unit 7183.
Charles Lewis Gazin Papers, Record Unit 7314.
Henry Helm Clayton Papers, Record Unit 7153.
Lucille Quarry Mann, Record Unit 9513.
National Zoological Park Papers, 1887–1965 and undated, Record Unit 74.
Office of the Secretary, Records, 1890–1929, Record Unit 45.
Office of the Secretary, Records, 1925–1949, Record Unit 46.
Paul H. Oehser Papers, Record Unit 9507.
Science Service Records, Record Unit 7091.
Science Service Engineering News Morgue Files, Accession 06-105.
Science Service Physics News Morgue Files, Accession 06-134.
Science Service Records, 1920s–1970s, Accession 90-105.
Science Service Records, 1925–1966, Accession 97-020.
Smithsonian Institution Press, Records, 1914–1965, Record Unit 83.
United States National Museum, Department of Engineering and Industries, Records, 1891–1959, Record Unit 84.
United States National Museum, Division of Vertebrate Paleontology, Records, circa 1889–1957, Record Unit 156.
Webster P. True Papers, 1914–1972, T91027.
William M. Mann and Lucille Quarry Mann Papers, circa 1885–1981, Record Unit 7293.

Archives Center, National Museum of American History, Smithsonian Institution

"Adventures in Science" Radio Programs, 1955–1956.

Curatorial Collections, National Museum of American History, Smithsonian Institution

Chemistry Collections, Division of Medicine and Science.
Engineering Collections, Division of Work and Industry.

National Museum of Natural History, Smithsonian Institution

Austin Hobart Clark Papers in the curatorial collections of Dr. David Pawson (to be transferred to the Smithsonian Institution Archives).

Library of Congress, Manuscript Collections

American Psychological Association Records.
Benjamin C. and Sidonie Matsner Gruenberg Papers, MSS 24265.
George Ellery Hale Papers, 1882-1937, Carnegie Institution of Washington (microfilm).
Papers of Roy Wilson Howard, 1911-1966, MSS 26583.
James McKeen Cattell Papers, MSS 15412.
John C. Merriam Papers, MSS 32706.
The Robert Andrews Millikan Collection at the California Institute of Technology (microfilm).

Hagley Museum and Library

Cavalcade of America Collection.

U.S. Fish and Wildlife Service, National Conservation and Training Center Archives

National Wildlife Federation Collection.

Milton S. Eisenhower Library, Johns Hopkins University

Johns Hopkins University, Office of News and Information Services, Records, 1946-(ongoing), Record Group 10.020.
Lynn Poole Papers, 1948-1976, MS 27, Special Collections.

National Academies Archives

Central Policy Files, 1919-1923.
Central Policy Files, 1923-1931.

SCRIPTS

Scripts for many Smithsonian Radio Talks of the 1920s are located in the Smithsonian Institution Archives, especially Record Units 45 and 7183.
Scripts for *Science Snapshots* are in Smithsonian Institution Archives, Science Service Records, Box 112.
Scripts for *Cavalcade of America* are located in the Cavalcade of America Collection at the Hagley Museum and Library.
Scripts for some *Adventures in Research* broadcasts are in Smithsonian Institution Press, Records, 1914-1965, Record Unit 83, Box 17, Smithsonian Institution Archives.
A significant number of *Adventures in Science* scripts are in Science Service Records, Record Unit 7091, Series 10; additional scripts are located throughout other Science Service collections in the Smithsonian Institution Archives and in curatorial collections within various Smithsonian Institution museums.
Scripts for *The World Is Yours* are located in various Smithsonian Institution Archives collections, especially those related to the Office of the Secretary, Smithsonian Institution Press, and Webster P. True.

BOOKS

Annual Report of the Board of Regents of the Smithsonian Institution, 1919. Washington, D.C.: U.S. Government Printing Office, 1921.

Annual Report of the Board of Regents of the Smithsonian Institution, 1920. Washington, D.C.: U.S. Government Printing Office, 1922.

Annual Report of the Board of Regents of the Smithsonian Institution, 1925. Washington, D.C.: U.S. Government Printing Office, 1926.

Annual Report of the Board of Regents of the Smithsonian Institution, 1926. Washington, D.C.: U.S. Government Printing Office, 1927.

Annual Report of the Board of Regents of the Smithsonian Institution, 1927. Washington, D.C.: U.S. Government Printing Office, 1928.

Badash, Lawrence. *Scientists and the Development of Nuclear Weapons: From Fission to the Limited Test Ban Treaty, 1939-1963.* Atlantic Highland: Humanities Press, 1995.

Bailey, Robert Lee. *An Examination of Prime Time Network Television Special Programs, 1948 to 1966.* New York: Arno Press, 1979.

Baker, John C. *Farm Broadcasting: The First Sixty Years.* Ames: Iowa State University Press, 1981.

Baldesty, Gerald J. *E. W. Scripps and the Business of Newspapers.* Urbana: University of Illinois Press, 1999.

Barfield, Ray. *Listening to the Radio, 1920-1950.* Westport: Praeger, 1996.

Barnouw, Erik. *The Golden Web: A History of Broadcasting in the United States, 1933-1953.* New York: Oxford University Press, 1968.

————.*The Image Empire: A History of Broadcasting in the United States from 1953.* New York: Oxford University Press, 1970.

————. *The Sponsor: Notes on a Modern Potentate.* New York: Oxford University Press, 1978.

————. *A Tower in Babel: A History of Broadcasting in the United States to 1933.* New York: Oxford University Press, 1966.

————., ed. *Radio Drama in Action: Twenty-Five Plays of a Changing World.* New York: Farrar and Rinehart, 1945.

Bauer, W. W., and Leslie Edgley. *Your Health Dramatized: Selected Radio Scripts.* Philadelphia: W. B. Saunders, 1937.

Bauer, W. W., and Thomas G. Hall. *Health Education of the Public: A Practical Manual of Technic.* Philadelphia: W. B. Saunders, 1937.

Baughman, James L. *The Republic of Mass Culture: Journalism, Filmmaking, and Broadcasting in America Since 1941.* Baltimore: Johns Hopkins University Press, 1992.

Beebe, William. *Half Mile Down.* New York: Duell, Sloan and Pearce, 1951.

Bingham, Walter V., ed. *Psychology Today: Lectures and Study Manual.* Chicago: University of Chicago Press, 1932.

Bird, William L., Jr. *"Better Living": Advertising, Media, and the New Vocabulary of Business Leadership, 1935-1955.* Evanston: Northwestern University Press, 1999.

Bluem, A. William. *Documentary in American Television: Form, Function, Method.* New York: Hastings House, 1965.

Boyer, Paul. *By the Bomb's Early Light: American Thought and Culture at the Dawn of the Atomic Age.* New York: Pantheon, 1985.

————. *Fallout: A Historian Reflects on America's Half-Century Encounter with Nuclear Weapons.* Columbus: Ohio State University Press, 1998.

————. *Purity in Print: Book Censorship in America from the Gilded Age to the Computer Age.* 2nd ed. Madison: University of Wisconsin Press, 2002.

Brians, Paul. *Nuclear Holocausts: Atomic War in Fiction, 1895-1984.* Kent, Ohio: Kent State University Press, 1987.

Brooks, Tim, and Earle Marsh. *The Complete Directory to Prime Time Network and Cable TV Shows, 1946–Present.* 6th ed. New York: Ballantine Books, 1995.

Brown, Robert J. *Manipulating the Ether: The Power of Broadcast Radio in Thirties America.* Jefferson, N.C.: McFarland, 1998.

Buckingham, David. *Children Talking Television: The Making of Television Literacy.* London: Falmer Press, 1993.

Burgess, Thornton W. *Now I Remember: Autobiography of an Amateur Naturalist.* Boston: Little, Brown, 1960.

Burlingame, Roger. *Endless Frontiers: The Story of McGraw-Hill.* New York: McGraw-Hill, 1959.

Burnham, John C. *How Superstition Won and Science Lost: Popularizing Science and Health in the United States.* New Brunswick: Rutgers University Press, 1987.

Buxton, Frank, and Bill Owen. *The Big Broadcast, 1920–1950.* Lanham, Md.: Scarecrow Press, 1997.

Cantril, Hadley. *The Invasion from Mars: A Study in the Psychology of Panic.* Princeton: Princeton University Press, 1940.

———. *Public Opinion 1935–1946.* Princeton: Princeton University Press, 1951.

Cater, Douglass. *The Fourth Branch of Government.* Boston: Houghton Mifflin, 1959.

Charnley, Mitchell V. *News by Radio.* New York: Macmillan, 1948.

Columbia Broadcasting System. *Crescendo: A Chronicle of Radio's Unique Power to Move People to Direct Action.* New York: CBS, 1947.

Columbia University Bureau of Applied Social Research. *The People Look at Radio.* Chapel Hill: University of North Carolina Press, 1946.

Craig, Douglas B. *Fireside Politics: Radio and Political Culture in the United States, 1920–1940.* Baltimore: Johns Hopkins University Press, 2000.

Crisell, Andrew. *Understanding Radio.* London: Methuen, 1986.

Davidson, Keay. *Carl Sagan: A Life.* New York: John Wiley and Sons, 1999.

Davis, Jeffrey. *Children's Television, 1947–1990.* Jefferson, N.C.: McFarland, 1995.

Davis, Watson. *The Advance of Science.* New York: Doubleday Doran, 1934.

Douglas, Susan. *Listening In: Radio and the American Imagination.* New York: Times Books, 1999.

Draper, Benjamin, ed. *The "Science in Action" TV Library,* Vol. 1. New York: Merlin Press, 1956.

Dunning, John. *On the Air: The Encyclopedia of Old-Time Radio.* New York: Oxford University Press, 1998.

———. *Tune In Yesterday: The Ultimate Encyclopedia of Old-Time Radio, 1925–1976.* New York: Prentice-Hall, 1976.

Dupree, A. Hunter. *Science and the Federal Government: A History of Policies and Activities to 1940.* Rev. ed. Baltimore: Johns Hopkins University Press, 1986.

Dupuy, Judy. *Television Show Business.* Schenectady: General Electric, 1945.

Einstein, Daniel. *Special Edition: A Guide to Network Television Documentary Series and Special News Reports, 1955–1979.* Metuchen, N.J.: Scarecrow Press, 1987.

Eisenstein, Elizabeth L. *The Printing Press as an Agent of Change.* 2 vols. Cambridge: Cambridge University Press, 1979.

Erikson, Hal. *Syndicated Television: The First Forty Years, 1947–1987.* Jefferson, N.C.: McFarland, 1989.

Ettema, James S., and D. Charles Whitney. *Audiencemaking: How the Media Create the Audience.* Thousand Oaks: Sage Publications, 1994.

Fielding, Raymond. *The American Newsreel, 1911–1967.* Norman: University of Oklahoma Press, 1972.

———. *The March of Time, 1935–1951.* New York: Oxford University Press, 1978.

Fowler, Gene, and Bill Crawford. *Border Radio.* Austin: Texas Monthly Press, 1987.

Gabler, Neal. *Life the Movie: How Entertainment Conquered Reality.* New York: Knopf, 1998.

————. *Walt Disney: The Triumph of the American Imagination*. New York: Knopf, 2006.

Gamson, Joshua. *Claims to Fame: Celebrity in Contemporary America*. Berkeley: University of California Press, 1994.

Geier, Leo. *Ten Years with Television at Johns Hopkins*. Baltimore: Johns Hopkins University, 1958.

Gianakos, Larry. *Television Drama Series Programming: A Comprehensive Chronicle, 1947–59*. Lanham, Md.: Scarecrow Press, 1980.

————. *Television Drama Series Programming: A Comprehensive Chronicle, 1959–75*. Lanham, Md.: Scarecrow Press, 1978.

Gilbert, James Burkhart. *Redeeming Culture: American Religion in an Age of Science*. Chicago: University of Chicago Press, 1997.

Godfrey, Donald G., and Frederic A. Leigh, eds. *Historical Dictionary of American Radio*. Westport: Greenwood Press, 1998.

Goodell, Rae. *The Visible Scientists*. Boston: Little, Brown, 1977.

Gorman, Paul R. *Left Intellectuals and Popular Culture in Twentieth-Century America*. Chapel Hill: University of North Carolina Press, 1996.

Gould, Carol Grant. *The Remarkable Life of William Beebe: Explorer and Naturalist*. Washington, D.C.: Island Press/Shearwater Books, 2004.

Grams, Martin, Jr. *The Official Guide to the History of the Cavalcade of America*. Self-published, 1998.

Gruenberg, Benjamin C. *Science and the Public Mind*. New York: McGraw-Hill, 1935.

Haber, Heinz. *Our Friend the Atom*. New York: Simon & Schuster, 1957.

Halper, Donna L. *Invisible Stars: A Social History of Women in American Broadcasting*. Armonk: M. E. Sharpe, 2001.

Hance, Robert T. *A Series of Eight Radio Talks on Zoology, Old and New (with Select Bibliography)*. Radio Publication no. 39. Pittsburgh: University of Pittsburgh, 1928.

Handel, Leo A. *Hollywood Looks at Its Audience: A Report of Film Audience Research*. Urbana: University of Illinois Press, 1950.

Hawes, William. *American Television Drama: The Experimental Years*. University: University of Alabama Press, 1986.

Hawkins, Reginald R. *Technical Books and the War: An Exhibition*. New York: New York Public Library, 1943.

Haynes, Rosalynn D. *From Faust to Strangelove: Representations of the Scientist in Western Literature*. Baltimore: Johns Hopkins University Press, 1994.

Head, Sydney W. *Broadcasting in America: Survey of Television and Radio*. Cambridge: Riverside Press, 1956.

Heldenfels, R. D. *Television's Greatest Year—1954*. New York: Continuum, 1994.

Hester, Harriet H., H. L. Fishel, and Martin Magner. *Television in Health Education*. Chicago: American Medical Association, 1955.

Hickerson, Jay. *The New, Revised Ultimate History of Network Radio Programming and Guide to All Circulating Shows*. Jefferson, N.C.: McFarland, 1996.

Hilmes, Michelle. *Only Connect: A Cultural History of Broadcasting in the United States*. Belmont: Wadsworth, 2001.

————. *Radio Voices: American Broadcasting, 1922–1952*. Minneapolis: University of Minnesota Press, 1997.

Himmelstein, Hal. *Television Myth and the American Mind*. 2nd ed. New York: Praeger, 1994.

Horten, Gerd. *Radio Goes to War: The Cultural Politics of Propaganda during World War II*. Berkeley: University of California Press, 2002.

Joerg, W. L. G. *The Work of the Byrd Antarctic Expedition, 1928–1930*. New York: American Geographical Society, 1930.

Jowett, Garth. *Film, The Democratic Art*. Boston: Focal Press, 1985.

Kevles, Daniel J. *The Physicists: The History of a Scientific Community in Modern America*. New York: Alfred A. Knopf, 1978.

Key, V. O., Jr. *Public Opinion and American Democracy*. New York: Alfred A. Knopf, 1961.

Kiernan, Vince. *Embargoed Science*. Urbana: University of Illinois Press, 2006.

Kisseloff, Jeff. *The Box: An Oral History of Television, 1920-1961*. New York: Viking, 1995.

Kohlstedt, Sally Gregory, Michael M. Sokal, and Bruce V. Lewenstein. *The Establishment of Science in America*. New Brunswick: Rutgers University Press, 1999.

Koop, Theodore Frederick. *Weapon of Silence*. Chicago: University of Chicago Press, 1946.

LaFollette, Marcel C. *Making Science Our Own: Public Images of Science, 1910-1955*. Chicago: University of Chicago Press, 1990.

Landry, Robert J. *Who, What, Why Is Radio?* New York: George W. Stewart, 1942.

Lazarsfeld, Paul F. *Radio and the Printed Page*. 1940; New York: Arno Press, 1971.

Lazarsfeld, Paul F., and Patricia L. Kendall. *Radio Listening in America: The People Look at Radio—Again*. New York: Prentice-Hall, 1948.

Lazarsfeld, Paul F., and Frank N. Stanton, eds. *Communications Research, 1948-1949*. New York: Harper & Brothers, 1949.

Lears, T. Jackson. *Fables of Abundance: A Cultural History of Advertising in America*. New York: Basic Books, 1994.

Lee, R. Alton. *The Bizarre Careers of John R. Brinkley*. Lexington: University Press of Kentucky, 2002.

Lewis, Tom. *Empire of the Air: The Men Who Made Radio*. New York: Edward Burlingame, 1991.

Lovell, Russell A., Jr. *The Cape Cod Story of Thornton W. Burgess*. Town of Sandwich, Massachusetts, 1974.

Lutts, Ralph H. *The Nature Fakers: Wildlife, Science and Sentiment*. Golden: Fulcrum Publishing, 1990.

Lynes, Russell. *The Lively Audience: A Social History of the Visual and Performing Arts in America, 1890-1950*. New York: Harper & Row, 1985.

Marchand, Roland. *Advertising the American Dream: Making Way for Modernity, 1920-1940*. Berkeley: University of California Press, 1985.

Marsh, C. S., ed. *Educational Broadcasting 1936, Proceedings of the First National Conference on Educational Broadcasting, held in Washington, D.C., on December 10, 11, and 12, 1936*. Chicago: University of Chicago Press, 1937.

Marshall, Roy K. *The Nature of Things*. New York: Henry Holt, 1951.

Marzolf, Marion. *Up from the Footnote: A History of Women Journalists*. New York: Hastings House, 1977.

McDonald, J. Fred. *Don't Touch That Dial! Radio Programming in American Life, 1920-1960*. Chicago: Nelson-Hall, 1979.

McLuhan, Marshall. *Understanding Media: The Extensions of Man*. New York: McGraw-Hill, 1964.

McNeil, Alex. *Total Television*. 4th ed. New York: Penguin Books, 1996.

Meigs, Frances B. *My Grandfather, Thornton W. Burgess: An Intimate Portrait*. Beverly: Commonwealth Editions, 1998.

Melody, William. *Children's Television: The Economics of Exploitation*. New Haven: Yale University Press, 1973.

Mighetto, Lisa. *Wild Animals and American Environmental Ethics*. Tucson: University of Arizona Press, 1991.

Millikan, Robert A. *Reprint of His Address in May 1931 to the First National Assembly of the National Advisory Council on Radio in Education*. Chicago: University of Chicago Press, 1931.

Mitman, Gregg. *Reel Nature: America's Romance with Wildlife on Film*. Cambridge: Harvard University Press, 1999.

Mott, Frank Luther. *Journalism in Wartime: The University of Missouri's Thirty-Fourth Annual "Journalism Week" in Print.* 1943; Westport: Greenwood Press, 1984.

National Advisory Council on Radio in Education (NACRE). *Psychology Today.* Listener Notebook no. 1, prepared by Henry E. Garrett and Walter V. Bingham. New York: NACRE, 1931.

———. *Radio and Education: Proceedings of the First Assembly of the National Advisory Council on Radio in Education.* Chicago: University of Chicago Press, 1931.

———. *Radio and Education: Proceedings of the Second Annual Assembly of the National Advisory Council on Radio in Education.* Chicago: University of Chicago Press, 1932.

———. *Radio and Education: Proceedings of the Third Annual Assembly of the National Advisory Council on Radio in Education.* Chicago: University of Chicago Press, 1933.

———. *Radio and Education: Proceedings of the Fourth Annual Assembly of the National Advisory Council on Radio in Education.* Chicago: University of Chicago Press, 1934.

National Association of Educational Broadcasters (NAEB). *Lincoln Lodge Seminar on Educational Television: Proceedings.* Urbana: NAEB, 1953.

Nelkin, Dorothy. *Selling Science.* New York: W. H. Freeman, 1987.

Neuman, Russell. *The Future of the Mass Audience.* New York: Cambridge University Press, 1991.

New York Academy of Medicine. *Radio in Health Education.* New York: Columbia University Press, 1945.

Nye, Russel. *The Unembarrassed Muse: The Popular Arts in America.* New York: Dial Press, 1970.

Ohmann, Richard. *Selling Culture: Magazines, Markets, and Class at the Turn of the Century.* London: Verso, 1996.

Pack, Arthur Newton, and E. Laurence Palmer. *The Nature Almanac: A Handbook of Nature Education.* Washington, D.C.: American Nature Association, 1927.

Poole, Lynn. *Science via Television.* Baltimore: Johns Hopkins University Press, 1950.

Postman, Neil. *Amusing Ourselves to Death: Public Discourse in the Age of Show Business.* New York: Penguin Books, 1985.

Poundstone, William. *Carl Sagan: A Life in the Cosmos.* New York: Henry Holt, 1999.

President's Research Committee on Social Trends. *Recent Social Trends in the United States.* 2 vols. New York: McGraw-Hill, 1933.

Price, Don K. *The Scientific Estate.* Cambridge: Harvard University Press/Belknap Press, 1965.

Real, Michael R. *Mass-Mediated Culture.* Englewood Cliffs: Prentice-Hall, 1977.

Relyea, Harold C., ed. *Striking a Balance: National Security and Scientific Freedom.* Washington, D.C.: American Association for the Advancement of Science, 1985.

A Resume of CBS Broadcasting Activities during 1937. New York: CBS, 1938.

Reynolds, Neil B., and Ellis L. Manning, eds. *Excursions in Science.* 1939; Freeport: Books for Libraries Press, 1972.

Rose, Brian G., ed. *TV Genres: A Handbook and Reference Guide.* New York: Greenwood Press, 1985.

Sayre, Jeanette. *An Analysis of the Radiobroadcasting Activities of Federal Agencies.* Studies in the Control of Radio no. 3. Cambridge: Harvard University, 1941.

Scheuer, Jeffrey. *The Sound Bite Society: Television and the American Mind.* New York: Four Walls Eight Windows Publishers, 1999.

Schickel, Richard. *Intimate Strangers: The Culture of Celebrity in America.* 1985; Chicago: Ivan R. Dee, 2000.

Schwartz, Tony. *The Responsive Chord.* New York: Anchor Books, 1973.

Serving through Science, A Series of Talks Delivered by American Scientists on the New York Philharmonic-Symphony Program. New York: U.S. Rubber Company, 1946.

Shapley, Harlow, and Cecilia H. Payne, eds. *The Universe of Stars: Radio Talks from the Harvard Observatory.* Cambridge: Observatory, 1926. Rev. ed., 1929.

Shayon, Robert Lewis. *Operation Crossroads as Broadcast from the Coolidge Auditorium of the Library of Congress*. Washington, D.C.: U.S. Library of Congress, 1946.

Shurcliff, W. A. *Bombs at Bikini: The Official Report of Operation Crossroads*. New York: William H. Wise, 1947.

Silverman, Alexander, Charles G. King, Kendell S. Tesh, Alexander Lowy, Gebherd Stegeman, and Carl J. Engelder. *A Series of Seven Radio Talks on Chemistry and Human Progress*. Radio Publication no. 21. Pittsburgh: University of Pittsburgh, 1926.

Smith, Alice Kimball. *A Peril and A Hope: The Scientists' Movement in America, 1945-47*. Rev. ed. Cambridge: MIT Press, 1970.

Smith, Anthony. *The Shadow in the Cave: The Broadcaster, His Audience, and the State*. Urbana: University of Illinois Press, 1973.

————., ed. *Television: An International History*. Oxford: Oxford University Press, 1995.

Smithsonian Institution. *Report on the Progress and Condition of the United States National Museum for the Year Ended June 30, 1927*. Washington, D.C.: U.S. Government Printing Office, 1927.

Smulyan, Susan. *Selling Radio: The Commercialization of American Broadcasting, 1920-1934*. Washington, D.C.: Smithsonian Institution Press, 1994.

Sperber, A. M. *Murrow: His Life and Times*. New York: Freundlich Books, 1986.

Sterling, Christopher H., and John M. Kittross. *Stay Tuned: A Concise History of American Broadcasting*. Belmont: Wadsworth, 1978.

Stettel, Irving, ed. *Top TV Shows of the Year, 1954-1955*. New York: Hastings House, 1955.

Stokley, James, ed. *Science Marches On*. New York: Ives Washburn, 1951.

Strasser, Susan. *Satisfaction Guaranteed: The Making of the American Mass Market*. New York: Pantheon Books, 1989.

Streeter, Thomas. *Selling the Air: A Critique of the Policy of Commercial Broadcasting in the United States*. Chicago: University of Chicago Press, 1996.

Sullivan, Mark. *Our Times: The United States, 1900-1925*. Vol. 6, *The Twenties*. New York: Charles Scribner's Sons, 1935.

Summers, Harrison B., ed. *A Thirty-Year History of Programs Carried on National Radio Networks in the United States, 1926-1956*. 1958; New York: Arno Press and New York Times, 1971.

Summers, Robert E., ed. *Federal Information Controls in Peacetime*. New York: H. W. Wilson, 1949.

————., ed. *Wartime Censorship of Press and Radio*. New York: H. W. Wilson, 1942.

Swartz, Jon D., and Robert C. Reinehr. *Handbook of Old-Time Radio: A Comprehensive Guide to Golden Age Radio Listening and Collecting*. Metuchen, N.J.: Scarecrow Press, 1993.

Sweeney, Michael S. *Secrets of Victory: The Office of Censorship and the American Press and Radio in World War II*. Chapel Hill: University of North Carolina Press, 2001.

Taylor, Glenhall. *Before Television: The Radio Years*. South Brunswick and New York: A. S. Barnes, 1979.

Terrace, Vincent. *The Complete Encyclopedia of Television Programs 1947-1976*. 2 vols. South Brunswick and New York: A. S. Barnes, 1976.

————. *Radio Programs, 1924-1984: A Catalog of Over 1800 Shows*. Jefferson, N.C.: McFarland, 1999.

Thomas, Erwin K., and Brown H. Carpenter, eds. *Handbook on Mass Media in the United States*. Westport: Greenwood Press, 1994.

Thompson, James Stacey. *The Technical Book Publisher in Wartimes*. New York: New York Public Library, 1942.

Tobey, Ronald C. *The American Ideology of National Science, 1919-1930*. Pittsburgh: University of Pittsburgh Press, 1971.

Toumey, Christopher P. *Conjuring Science: Scientific Symbols and Cultural Meanings in American Life*. New Brunswick: Rutgers University Press, 1996.

Triangle Publications, Radio and Television Division. *The University of the Air: Commercial Television's Pioneer Effort in Education*. Philadelphia: Triangle Publications, 1959.

Trimble, Vance H. *The Astonishing Mr. Scripps: The Turbulent Life of America's Penny Press Lord*. Ames: Iowa State University Press, 1992.

True, Webster Prentiss. *The Smithsonian Institution*. New York: Series Publishers, 1949.

Turow, Joseph. *Media Industries: The Production of News and Entertainment*. New York: Longman, 1984.

———. *Media Systems in Society: Understanding Industries, Strategies and Power*. 2nd ed. New York: Longman, 1997.

Tyson, Levering. *Education Tunes In: A Study of Radio Broadcasting in Adult Education*. New York: American Association for Adult Education, 1930.

———, ed. *Radio and Education, Proceedings of the First National Assembly of the National Advisory Council on Radio in Education, 1931*. Chicago: University of Chicago Press, 1931.

———, ed. *Radio and Education, Proceedings of the Second Assembly of the National Advisory Council on Radio in Education, 1932*. Chicago: University of Chicago Press, 1932.

Tyson, Levering, and Josephine MacLatchy, eds. *Education on the Air... and Radio and Education 1935: Proceedings of the Sixth Annual Institute for Education by Radio... Combined with the Fifth Annual Assembly of the National Advisory Council on Radio in Education*. Chicago: University of Chicago Press, 1935.

Udelson, Joseph H. *The Great Television Race: A History of the American Television Industry, 1925-1941*. University: University of Alabama Press, 1982.

The University of Chicago Round Table. *America and the Atomic Age, Special Twentieth Anniversary Pamphlet, February 1931 to February 1951*. Chicago: University of Chicago, 1951.

von Schilling, James. *The Magic Window: American Television, 1939-1953*. New York: Haworth Press, 2003.

Waller, Judith C. *Broadcasting in the Public Service*. Chicago: John C. Swift, 1943.

———. *Radio: The Fifth Estate*. New York: Houghton Mifflin, 1946.

———. *Radio: The Fifth Estate*. 2nd ed. Boston: Houghton Mifflin, 1950.

Waples, Douglas, ed. *Print, Radio, and Film in a Democracy*. Chicago: University of Chicago Press, 1942.

Watts, Steven. *The Magic Kingdom: Walt Disney and the American Way of Life* Boston: Houghton Mifflin, 1997.

Weart, Spencer R. *Nuclear Fear: A History of Images*. Cambridge: Harvard University Press, 1988.

———. *Scientists in Power*. Cambridge: Harvard University Press, 1979.

Weart, Spencer R., and Gertrud Weiss Szilard. *Leo Szilard: His Version of the Facts, Selected Recollections and Correspondence*. Cambridge: MIT Press, 1978.

Weaver, Warren, ed. *The Scientists Speak*. New York: Boni & Gaer, 1947.

Welker, Robert Henry. *Natural Man: The Life of William Beebe*. Bloomington: Indiana University Press, 1975.

WGN: A Pictorial History. Chicago: WGN, 1961.

Williams, Raymond. *Television: Technology and Cultural Form*. New York: Schocken Books, 1975.

Wilson, Edmund. *American Earthquake: A Documentary of the Twenties and Thirties*. Garden City: Doubleday Anchor Books, 1958.

Winkler, Allan M. *Life under a Cloud: American Anxiety about the Atom*. New York: Oxford University Press, 1993.

Wylie, Max. *Best Broadcasts of 1938-39*. New York: Whittlesey House, 1939.

———. *Best Broadcasts of 1939-40*. New York: Whittlesey House, 1940.

———. *Best Broadcasts of 1940-41*. New York: Whittlesey House, 1941.

Yochelson, Ellis L. *Smithsonian Institution Secretary, Charles Doolittle Walcott*. Kent, Ohio: Kent State University Press, 2001.

OTHER PUBLISHED SOURCES

Alexander, George. "Don Herbert: Gee Wiz!" *SciQuest*, May/June 1981, 21–28.

"American Society of Newspaper Editors Reports on Atomic Information Problems." *Bulletin of the Atomic Scientists* 4 (July 1948): 211–12.

Amory, Cleveland. "Mr. Wizard." *TV Guide*, April 22, 1972, 48.

"Austin Hobert Clark." *Lepidopterists' News* 9 (1955): 151–56.

Badash, Lawrence, Elizabeth Hodes, and Adolph Tiddens. "Nuclear Fission: Reaction to the Discovery in 1939." *Proceedings of the American Philosophical Society* 130 (June 1986): 196–231.

Baker, Kenneth. "An Analysis of Radio's Programming." In *Communications Research, 1948–1949*, ed. Paul F. Lazarsfeld and Frank N. Stanton, 51–72. New York: Harper & Brothers, 1949.

Becker, Ron. "'Hear-and-See Radio' in the World of Tomorrow: RCA and the Presentation of Television at the World's Fair, 1939–1940." *Historical Journal of Film, Radio, and Television* 21 (October 2001): 361–78.

Bernays, Edward L. "Manipulating Public Opinion: The Why and the How." *American Journal of Sociology* 33 (May 1928): 958–71.

Beuick, Marshall D. "The Limited Social Effect of Radio Broadcasting." *American Journal of Sociology* 32 (January 1927): 615–22.

Bird, Winfred Wylam. "An Analysis of the Aims and Practice of the Principal Sponsors of Education by Radio in the United States." University of Washington Extension Series Bulletin no. 10 (August 1939).

Bryson, Lyman. "Can We Put Science on the Air?" *Journal of Educational Sociology* 14 (February 1941): 364–69.

"The Burgess Radio Nature League." *Literary Digest*, June 6, 1925, 30.

Butler, James J. "ASNE, Atomic Board Join in Security Study." *Editor & Publisher* 80 (December 6, 1947): 10.

Caldwell, Louis G. "Freedom of Speech and Radio Broadcasting." *Annals of the American Academy of Political and Social Science* 177 (January 1935): 179–207.

Clark, Austin H. "The American Association at St. Louis." *Scientific Monthly* 42 (February 1936): 186–88.

———. "The Atlantic City Meeting of the American Association for the Advancement of Science." *Scientific Monthly* 44 (February 1937): 186–91.

———. "Radio Talks." *Science* 71 (May 2, 1930): 454.

———. "Radio Talks." *Scientific Monthly* 35 (October 1932): 352–59.

———. "Science and the Press." *Science* 68 (August 3, 1928): 91–99.

———. "Science and the Radio." *Scientific Monthly* 34 (March 1932): 268–72.

———. "Selling Entomology." *Scientific Monthly* 32 (June 1931): 527–36.

———. "The Smithsonian Institution, Its Function and Its Future." *Science* 63 (February 1926): 147–57.

Cockerell, T. D. A. "Why Does Our Public Fail to Support Research?" *Scientific Monthly* 10 (April 1920): 368–71.

Crowther, Samuel. "Our Kings of Chemistry." *World's Work*, February 1921, 349.

Davies, Lawrence E. " 'Science in Action': Western Show Entertains and Gives Information." *New York Times*, July 5, 1953.

Davis, Watson. "Educational Television." *Scientific Monthly* 74 (June 1952): 366–67.

———. "Government Control of Radio Telephony." *Scientific Monthly* 14 (April 1922): 397.

———. "Science and the Press." *Annals of the American Academy of Political and Social Science* 219 (January 1942): 100–106.

———. "Science and the Press." *Vital Speeches* 2 (March 9, 1936): 361-62.

Davis, Watson, and Robert D. Potter. "Atomic Energy Released." *Science News Letter* 35 (February 11, 1939): 86-87.

De Pasquele, Sue. "Live from Baltimore—It's the Johns Hopkins Science Review!" *Johns Hopkins Magazine*, February 1995. http://www.jhu.edu/~jhumag/295web/scirevu.html (accessed May 3, 2006).

Denny, George V., Jr. "Radio Builds Democracy." *Journal of Educational Sociology* 14 (February 1941): 370-77.

Diamond, L. N. "Interpreting Science to the Public." *Scientific Monthly* 40 (April 1935): 370-75.

Dietz, David. "Science and the American Press." *Science* 85 (January 29, 1937): 107-12.

Dismuke, Diane. "Don 'Mr. Wizard' Herbert: Still a Science Trailblazer." *NEA Today* 12 (April 1994): 9.

Dryer, Sherman H. "The Dial Take the Hindmost." *Journal of Educational Sociology* 14 (February 1941): 524-32.

Fisher, Sterling. "The Radio and Public Opinion." *Public Opinion Quarterly* 2 (January 1938): 79-82.

Flynn, John T. "Edward L. Bernays: The Science of Ballyhoo." *Atlantic Monthly*, May 1932, 562-71.

Foust, James C. "E. W. Scripps and the Science Service." *Journalism History* 21 (Summer 1995): 58-64.

Frank, Glenn. "Radio as an Educational Force." *Annals of the American Academy of Political and Social Science* 177 (January 1935): 119-22.

Frost, Stanley. "Radio—Our Next Great Step Forward." *Collier's Weekly*, April 8, 1922, 3-4, 18-24.

"*General Electric Review:* How a Television Commercial Is Made (1956)." In *Mass Communications*, 2nd ed., ed. Wilbur Schramm, 161-74. Urbana: University of Illinois Press, 1960.

Gilbert, James. " 'Our Mr. Sun': Religion and Science in 50s America." *History Today* 45 (February 1995): 33-39.

Gitlin, Irving J. "Radio and Atomic-Energy Education." *Journal of Educational Sociology* 22 (January 1949): 327-30.

Goldsmith, H. H. "The Literature of Atomic Energy of the Past Decade." *Scientific Monthly* 68 (May 1949): 291-92.

Goodman, Mark. "The Radio Act of 1927 as a Product of Progressivism." *Media History Monographs* 2 (1998-1999).

Gould, Jack. "Science Offered Over Television." *New York Times*, October 18, 1950.

Groves, Leslie R. "People Should Learn about Nuclear Energy." *Journal of Educational Sociology* 22 (January 1949): 318-20.

Gruenberg, Benjamin C. "Science and the Layman." *Scientific Monthly* 40 (May 1935): 450-57.

Hamburger, Philip. "Experts." *New Yorker*, May 3, 1952, 68-70.

Harris, E. H. "Radio and the Press." *Annals of the American Academy of Political and Social Science* 177 (January 1935): 163-69.

Head, Sydney W. "Content Analysis of Television Drama Programs." *Quarterly of Film, Radio, and Television* 9 (Winter 1954): 175-94.

Hettinger, Herman S., ed. "New Horizons in Radio." Special issue. *Annals of the American Academy of Political and Social Science* 213 (January 1941).

"The Hopkins on TV." *Newsweek*, March 17, 1952, 88-89.

Kiernan, Vincent J. "Ingelfinger, Embargoes, and Other Controls on the Dissemination of Science News." *Science Communication* 18 (June 1997): 297-319.

Killeffer, D. H. "Chemical Education via Radio." *Journal of Chemical Education* 1 (March 1924): 43-48.

LaFollette, Marcel C. "Eyes on the Stars: Images of Women Scientists in Popular Culture." *Science, Technology, & Human Values* 13 (Fall 1988): 262-75.

———. "Science on Television: Influences and Strategies." *Daedalus* 111 (Fall 1982): 183-98.

———. "A Survey of Science Content in U.S. Radio Broadcasting, 1920s through 1940s: Scientists Speak in Their Own Voices." *Science Communication* 24 (September 2002): 4-33.

———. "A Survey of Science Content in U.S. Television Broadcasting, 1940s through 1950s: The Exploratory Years." *Science Communication* 24 (September 2002): 34-71.

———. "Taking Science to the Marketplace: Examples of Science Service's Presentation of Chemistry during the 1930s." *HYLE* 12 (2006): 67-97.

Leach, Eugene E. "Tuning Out Education: The Cooperative Doctrine in Radio" (originally published as "Snookered 50 Years Ago"). *Current*, December 13, 1999. http://www.current.org/coop.

Lewis, Tom. "'A Godlike Presence': The Impact of Radio on the 1920s and 1930s." *OAH Magazine of History* 6 (Spring 1992). http://www.oah.org/pubs/magazine/communication/lewis.html.

Lichty, Lawrence W. "Success Story." *Wilson Quarterly* 5 (Winter 1981): 63-64.

Livingston, Burton E. "The Permanent Secretary's Report on the Fifth Washington Meeting." *Science* 61 (February 6, 1925): 121-30.

Lumley, F. H. "Audiences and Educational Radio Talks." *Educational Radio Bulletin* 12 (January 11, 1933): 14-16.

"Lynn Poole, Won Early TV Prizes." *New York Times*, April 16, 1969.

Marshall, Roy K. "Televising Science." *Physics Today* 2 (January 1949): 26-30.

Marvin, C. F. "Radio Meteorological Services." *Annals of the American Academy of Political and Social Science* 142 (March 1929): 64-66.

McChesney, Robert W. "Media and Democracy: The Emergence of Commercial Broadcasting in the United States, 1927-1935." *OAH Magazine of History* 6 (Spring 1992). http://www.oah.org/pubs/magazine/communication.

Mechling, Elizabeth Walker, and Jay Mechling. "The Atom According to Disney." *Quarterly Journal of Speech* 81 (November 1995): 436-53.

Miller, Neville. "The Broadcaster Speaks." *Journal of Educational Sociology* 14 (February 1941): 323-24.

Morgan, Joy Elmer. "Asks States to Aid Education by Radio." *New York Times*, February 7, 1932.

Moulton, F. R. "Science by Radio." *Scientific Monthly* 47 (December 1938): 546-48.

———. "The Third Indianapolis Meeting of the American Association for the Advancement of Science and Its Associated Societies." *Science* 87 (February 4, 1938): 95-118.

"Nature League Proves Radio's Aid to Science." *Washington Post*, January 31, 1926.

Pawson, David, and Doris J. Vance. "Austin Hobart Clark (1880-1954): His Echinoderm Research and Contacts with Colleagues." In *12th International Echinoderm Conference, New Hampshire* (Lisse: Taylor & Francis, in press).

———. "On Board the *Albatross* in 1906: A Young Scientist Writes Home." Personal communication, 2007.

"The Permanent Secretary's Report on the Fifth Washington Meeting." *Science* 61 (February 6, 1925): 121-30.

Phillips, John C. "An Investigation of the Periodic Fluctuations in the Numbers of the Ruffed Grouse." *Science* 63 (January 22, 1926): 92-93.

Pigott, Leonard D. "Biology Reaches a New Horizon: Science on Television." *A.I.B.S. Bulletin* 1 (July 1951): 7.

Poole, Lynn. "The Challenge of Television." *College Art Journal* 8 (Summer 1949): 299-304.

"Radio Owners' Census Is Begun in District." *Washington Post*, December 2, 1923.

"Radio Talks from the Harvard College Observatory." *Science* 62 (November 13, 1925): 431.

Reid, Louis. "Broadcasting Books: Drama, Fiction, and Poetry on the Air." *Saturday Review of Literature* 19 (January 14, 1939): 14.

Relyea, Harold C. "Information, Secrecy, and Atomic Energy." *New York University Review of Law and Social Change* 10 (1980-1981): 265-86.

Rhees, David. "The Chemical Foundation and Popular Chemistry between the Wars." *Center for the History of Chemistry (CHOC) News* 3 (Spring 1985): 2-3.

————. "The Chemists' War: The Impact of World War I on the American Chemical Profession." *Bulletin of the History of Chemistry* 13-14 (1992-1993): 40-47.

Ridenour, Louis N. "Military Secrecy and the Atomic Bomb." *Fortune*, November 1945, 170-71.

Ritter, William E. "The Relation of E. W. Scripps to Science." *Science* 65 (March 25, 1927): 291-92.

————. "Science for the Millions." *Scientific American Monthly* 3 (May 1921): 462-64.

Russell, Doug. "Popularization and the Challenge to Science-Centrism in the 1930s." In *The Literature of Science: Perspectives on Popular Science Writing*, ed. Murdo William McRae, 37-53. Athens: University of Georgia Press, 1993.

Salisbury, Morse. "People Should Learn about Nuclear Energy." *Journal of Educational Sociology* 22 (January 1949): 322-23.

————. "Radio and the Farmer." *Annals of the American Academy of Political and Social Science* 177 (January 1935): 141-46.

Sarnoff, David. "Probable Influence of Television on Society." *Journal of Applied Physics* 10 (July 1939): 428.

————. "Radio." *Saturday Evening Post*, August 14, 1926.

Schatz, Willie. "Welcome Back, Mr. Wizard." *Washington Post TV Magazine*, November 9-15, 1980, 5.

Schulz, Gary. "Is Video a Good Education Medium?" *Wisconsin Alumnus* 53 (June 1952): 8-10.

"Science Hush-Hushed." *Time*, May 11, 1942, 90.

"Science Service Conference." *Science* 76 (August 19, 1932): 151-58.

"Science Service Conference. II." *Science* 76 (August 26, 1932): 180-84.

Siepmann, C. A. "Can Radio Educate?" *Journal of Educational Sociology* 14 (February 1941): 346-57.

Silverman, Milton. "Ma Bell's House of Magic." *Saturday Evening Post*, May 10, 1947, 15-17, 147.

Silverstone, Roger. "Science and the Media: The Case of Television." In *Images of Science: Scientific Practice and the Public*, ed. S. J. Doorman, 187-211. Aldershot: Gower, 1989.

Slosson, E. E. "Letter to the Editor." *Science* 55 (May 5, 1922): 480-82.

————. "A New Agency for the Popularization of Science." *Science* 53 (April 8, 1921): 321-23.

Sokal, Michael M. "From the Archives: Cattell and World War II Censorship." *Science, Technology & Human Values* 10 (Spring 1985): 24-27.

————. "Restrictions on Scientific Publication." *Science* 215 (March 5, 1982): 1182.

Sorenson, R. W. "World Is Now in Radio Age." *Los Angeles Times*, April 2, 1922.

Steinke, Jocelyn, and Marilee Long. "A Lab of Her Own? Portrayals of Female Characters on Children's Educational Science Programs." *Science Communication* 18 (December 1996): 91-115.

Studebaker, John W. "Promoting the Cause of Education by Radio." *Journal of Educational Sociology* 14 (February 1941): 325-33.

Suzuki, David T. "Information Overload: More TV Science Could Add to the Confusion." *SIPIscope* 14 (January/February 1986): 4.

Swain, Bruce M. "The Progressive, the Bomb, and the Papers." *Journalism Monographs*, no. 76 (May 1982).

Templeton, David. "Weird Science." *Sonoma County Independent*, September 23-29, 1999.

Thone, F. E. A., and E. W. Bailey. "William Emerson Ritter: Builder." *Scientific Monthly* 24 (March 1927): 256-62.

Vanocur, Sander. "TV's Failed Promise." *Center Magazine* 4 (November/December 1971): 44-50.

Ward, Henry B. "American Association for the Advancement of Science: The Joint Meeting at Rochester and Ithaca." *Science* 84 (July 31, 1936): 96-113.

―――. "The St. Louis Meeting of the American Association for the Advancement of Science and Associated Societies." *Science* 83 (February 7, 1936): 111-46.

Washburn, Patrick S. "The Office of Censorship's Attempt to Control Press Coverage of the Atomic Bomb During World War II." *Journalism Monographs*, no. 120 (April 1990).

"WBZ Starts Radio Nature Association." *Christian Science Monitor*, February 18, 1925.

Weart, Spencer R. "Scientists with a Secret." *Physics Today* 29 (February 1976): 23-30.

Wesolowski, James Walter. "Before Canon 35: WGN Broadcasts the Monkey Trial." *Journalism History* 2 (Fall 1975): 76-79, 86-87.

Yoder, Robert M. "TV's Shoestring Surprise." *Saturday Evening Post*, August 21, 1954, 30, 90-91.

UNPUBLISHED SOURCES

Bailey, Robert Lee. "An Examination of Prime Time Network Television Special Programs, 1948 to 1966." PhD diss., University of Wisconsin, 1967.

Ehrhardt, George Robert. "Descendants of Prometheus: Popular Science Writing in the United States, 1915-1948." PhD diss., Duke University, 1993.

National Association of Science Writers—Jane Stafford." Transcript of an interview conducted by Bruce V. Lewenstein, February 6, 1987, National Association of Science Writers Records, Cornell University Library.

Rhees, David J. "A New Voice for Science: Science Service under Edwin E. Slosson, 1921-1929." MA thesis, University of North Carolina, 1979.

Soranno, Alexander Mark. "A Descriptive Study of Television Network Prime Time Programming: 1958-59 through 1962-63." MA thesis, San Francisco State College, 1966.

Toon, Elizabeth A. "Managing the Conduct of the Individual Life: Public Health Education and American Public Health, 1910 to 1940." PhD diss., University of Pennsylvania, 1998.

Wakeley, Lillian Donley. "Adult Education about Atomic Energy, 1945-1948, as a Case Study in Science for Society." DEd diss., Pennsylvania State University, 1984.

WEB SITES

"Women of Science Service," Smithsonian Institution Archives Web exhibit http://siarchives.si.edu/research/sciservwomen.html.

VIDEO

Johns Hopkins Television Programs, 1948-1960, Videotape Collection, Special Collections, Milton S. Eisenhower Library, Johns Hopkins University.

Eight of the Bell science television specials (*Our Mr. Sun, Hemo the Magnificent, The Strange Case of the Cosmic Rays, The Unchained Goddess, Gateways to the Mind: The Story of the Human Senses, The Alphabet Conspiracy, The Thread of Life,* and *About Time*) are available in videotape and DVD collections.

Acknowledgments

THIS book rests on the sturdy foundations built by numerous scholars, and I gratefully acknowledge their influence on my work. Cultural historians like Erik Barnouw, Susan Douglas, Sydney Head, Michelle Hilmes, and Anthony Smith, for example, have analyzed the evolution of American broadcasting, explaining how Barnouw's "golden web" of radio responded to listeners, translated their preferences into advertising sales, and then spun dreams into a lucrative, powerful, and politically indispensable industry. During those same decades of the twentieth century, science was also transforming American culture and journalism, a change chronicled so well by John C. Burnham, Daniel J. Kevles, Ronald C. Tobey, Paul Boyer, Spencer R. Weart, James Gilbert, Bruce Lewenstein, and others. The late Dorothy Nelkin showed how to disentangle the connections between science and the public and advanced our understanding of how scientists respond to the media spotlight.

Science Service began as an experiment and evolved steadily into a success. The preservation of the organization's records took a more dramatic turn, however, shortly after the death of Watson Davis in 1967. The late Audrey Davis, then a newly appointed curator at the Smithsonian Institution, pushed in the late 1960s for acquisition of a major section of the editorial records of Science Service when these materials were about to be destroyed. As Watson's daughter-in-law Audrey was familiar with the organization's work, but as an historian of science she also recognized the potential importance of these records. For the history of science and science journalism, it was a fortuitous combination of knowledge and judgment. She lived long enough to learn of some of the treasures those dusty boxes held. Future researchers will uncover more. I hope that they too will remember and honor Audrey's foresight.   .

David Rhees had begun many years ago, with only some of these records available, to tell the story of Science Service. This book extends the road David began to build, but does not purport to tell the complete history of an organization that continues to operate today. I have focused on the radio broadcasts and the evolution of the Science Service approach to journalism and popularization. There is much more to tell about its science education activities, documentation and translation projects, magazines, cartoons, columns, and the contributions of individual staff members and stringers.

The majority of the Science Service Records eventually came to the Smithsonian Institution Archives (SIA). I will be forever grateful to all SIA staff, past and present, but shout out a special "thank you" to Ellen Alers, Tammy Peters, Jim Steed, Shawn Johnstone, and Tracey Robinson, who were always willing to retrieve "one more box" and (most important) kept me laughing all the way. You are the best. I also thank the archivists at the Library of Congress Manuscript Division and the Hagley Museum and Library; Janice Goldblum at the National Academies Archives; and Heidi Herr and other archivists at the Special Collections at Johns Hopkins University's Milton S. Eisenhower Library, for help in obtaining videotapes and photographs.

All of us—historians as well as the consumers of history—owe a substantial debt to archivists, conservators, and librarians. They literally do history's heavy lifting, applying intellectual, organizational, and (as needed) physical ingenuity to assure that original records will survive into the future. In the twenty-first century, however, archives, collections, and libraries are increasingly under duress—underfunded, understaffed, and underappreciated by their patrons and (most tragically) their host institutions. Preservation of these facilities is essential if, in the future, we are, as Watson Davis urged, to "keep faith with truth."

THIS book has taken a circuitous journey toward publication, and I owe considerable thanks to people who commented on earlier versions or related work, especially Lawrence Badash, Mary Jane McKinven, and Carol Rogers. Jonathan Cobb and Marc Rothenberg took time to provide feedback in the later stages. The anonymous referees provided invaluable comments and much-needed encouragement. The staff at the University of Chicago Press, especially Christie Henry and Joel Score, have been supportive, professional, and enthusiastic—everything a weary author needs to edit a manuscript into submission. Thank you to you all.

I have benefited through the years from help by friends and colleagues throughout the Smithsonian Institution, but owe special thanks to Ann Seeger and Pete Daniel at the National Museum of American History. And I owe a tremendous debt to David Pawson and Doris Vance of the National Museum of Natural History, who graciously made available a newly acquired treasure trove of papers related to Austin H. Clark and thereby unlocked my understanding of Clark's relationship with Thornton W. Burgess. Your kindnesses have been extraordinary.

The support of many friends, such as Mary Cole, Lisa Helperin, Jonathan Coopersmith, Jane and Barney Finn, Don Reisman, Jack White, and Chris Foreman, has been unwavering.

Chris Cherniak, you get special thanks because you helped me find my voice.

To Irene and Mickey Schubert, may there be many more Nats baseball games and Thanksgivings together. As Frank Thone said, "New Turkeys for Old!"

To Margaret Vining and Bart Hacker, this book has been calorically enriched by your culinary delights, conversations, comfort, and company. My turn to cook the greens.

To David R. Gessner, you have been there in thick and thin, sorrow and joy. I sure hope you like this broadcast.

To BBW, my little Mozart and Mick Jagger fan, I say, "Turn the radio on." Your songs echo through these pages.

To my husband Jeffrey Stine, you have kept me happy, steady on the course, reinforced, revived, enthused, inspired, focused, and, above all, laughing whenever the clouds rolled in. This one's dedicated to you.

Marcel C. LaFollette
May 2007

Index